ECOLOGY AND THEOLOGY IN THE
ANCIENT WORLD

Also published by Bloomsbury

GREEK AND ROMAN CALENDARS
Robert Hannah

REFLECTIONS AND NEW PERSPECTIVES ON VIRGIL'S GEORGICS
Nicholas Freer & Bobby Xinyue

ROME AND THE LITERATURE OF GARDENS
Victoria Emma Pagàn

THE BOUNDARIES OF ART AND SOCIAL SPACE IN ROME
Frederick Jones

THE POLITICS OF OLYMPUS
Jenny Strauss Clay

VIRGIL'S GARDEN
Frederick Jones

ECOLOGY AND THEOLOGY IN THE ANCIENT WORLD

CROSS-DISCIPLINARY PERSPECTIVES

Edited by Ailsa Hunt and Hilary Marlow

BLOOMSBURY ACADEMIC
LONDON • NEW YORK • OXFORD • NEW DELHI • SYDNEY

BLOOMSBURY ACADEMIC
Bloomsbury Publishing Plc
50 Bedford Square, London, WC1B 3DP, UK
1385 Broadway, New York, NY 10018, USA

BLOOMSBURY, BLOOMSBURY ACADEMIC and the Diana logo are trademarks of
Bloomsbury Publishing Plc

First published in Great Britain 2019
Paperback edition published 2021

Copyright © Ailsa Hunt, Hilary Marlow and Contributors, 2019

Ailsa Hunt and Hilary Marlow have asserted their right under the Copyright,
Designs and Patents Act, 1988, to be identified as Authors of this work.

For legal purposes the Preface on p. vi constitutes an extension of this copyright page.

Cover design: Terry Woodley
Cover image: Wall decorated with third style fresco depicting garden and birds.
From Pompeii, 1st century AD. (Photo By DEA/L. PEDICINI/De Agostini/Getty Images)

All rights reserved. No part of this publication may be reproduced or transmitted in any form or
by any means, electronic or mechanical, including photocopying, recording, or any information
storage or retrieval system, without prior permission in writing from the publishers.

Bloomsbury Publishing Plc does not have any control over, or responsibility for, any third-party
websites referred to or in this book. All internet addresses given in this book were correct at the
time of going to press. The author and publisher regret any inconvenience caused if addresses
have changed or sites have ceased to exist, but can accept no responsibility for any such changes.

A catalogue record for this book is available from the British Library.

Library of Congress Cataloging-in-Publication Data
Names: Hunt, Ailsa, editor.
Title: Ecology and theology in the ancient world : cross-disciplinary
perspectives / edited by Ailsa Hunt & Hilary Marlow.
Description: New York : Bloomsbury Academic, 2019. | Includes bibliographical
references and index.
Identifiers: LCCN 2018040920| ISBN 9781350004047 (hardback) | ISBN
9781350004054 (epub)
Subjects: LCSH: Ecology--Religious aspects. | Ecotheology.
Classification: LCC BL65.E36 E365 2019 | DDC 201/.77093--dc23 LC record available at
https://lccn.loc.gov/2018040920

ISBN: HB: 978-1-3500-0404-7
        PB: 978-1-3501-8328-5
        ePDF: 978-1-3500-0406-1
        eBook: 978-1-3500-0405-4

Typeset by Integra Software Services Pvt. Ltd.

To find out more about our authors and books visit www.bloomsbury.com and sign up for our
newsletters.

CONTENTS

Preface   vi
List of Abbreviations   vii
Notes on Contributors   x

Introduction   Ailsa Hunt and Hilary Marlow   1

1   **Ancient Ideas of Politics: Mediating between Ecology and Theology**   Melissa Lane   13

2   **The Ecology of the *Sibylline Oracles***   Helen Van Noorden   25

3   **Self-Sufficiency as a Divine Attribute in Greek Philosophy**   David Sedley   41

4   **A Lighter Shade of Green: Stoic Gods and Environmental Virtue Ethics**   Christoph Jedan   49

5   **Cosmic Beauty in Stoicism: A Foundation for an Environmental Ethic as Love of the Other?**   Jula Wildberger   63

6   **Some Ancient Philosophical and Religious Roots of Modern Environmentalism**   Robin Attfield   75

7   **Creatures in Creation: Human Perceptions of the Sea in the Hebrew Bible in Ecological Perspective**   Rebecca Watson   91

8   **Reconsidering the Chthonic in Aeschylus' *Oresteia*: Erinyes, the Earth's Resources and the Cosmic Order**   Emmanuela Bakola   103

9   **The Anguish of the Earth: Ecology and Warfare in the First World War and the Bible**   Hilary Marlow   119

10   **Pagan Animism: A Modern Myth for a Green Age**   Ailsa Hunt   137

Notes   153
References   176
Index   195

PREFACE

This collection of essays is the fruit of a conscious decision to bridge traditional boundaries and invite collaboration, both between disciplines in contemporary academia, and also across the temporal distance between the ancient world and contemporary concerns. The chapters of the book began life as papers presented at a conference 'Greening the Gods: Ecology and Theology in the Ancient World', held in Cambridge in March 2014.

We are grateful to the Faculty of Classics, University of Cambridge for a grant that enabled the conference to take place and to The Faraday Institute for Science and Religion for their administrative and financial support in hosting the conference at St Edmund's College, Cambridge. We would like to express our thanks to the contributors to the volume for their thought-provoking work, commitment to the volume and patience, especially as we finalized the manuscript. Finally, we are grateful to Bloomsbury for agreeing to publish this volume and especially to Alice Wright and Emma Payne for their help and support throughout the publication process.

Ailsa Hunt
Hilary Marlow
Aug 2018

ABBREVIATIONS

General abbreviations

AB	Anchor Bible
ANET	Pritchard, J. B (1969), *Ancient Near Eastern Texts Relating to the Old Testament*, 3rd edn, Princeton, NJ: Princeton University Press
BCE	Before Common Era
BibOr	Biblica et orientalia
BZAW	Beihefte zur Zeitschrift für die alttestamentliche Wissenschaft
CCJ	Cambridge Classical Journal
CE	Common Era
CIL	Corpus Inscriptionum Latinarum
CPh	Classical Philology
DCH	Clines, D. (1993), *The Dictionary of Classical Hebrew*, Sheffield: Sheffield Academic Press
DK	Diels-Kranz. (Diels, H., revised by Kranz, W. (1952) *Die Fragmente der Vorsokratiker*. Berlin: Weidmann)
EA	El Amarna letters (J.A. Knudtzon, (1915) *Die el-Amarna Tafeln*. Leipzig: J. C. Hinrichs)
frg	Fragment
HAT	Handbuch zum Alten Testamentum
HSM	Harvard Semitic Monographs
ICS	Illinois Classical Studies
JBL	Journal of Biblical Literature
JNES	Journal of Near Eastern Studies
JPSA	Jewish Publication Society of America
KJV	King James Version
KTU	M. Dietrich, O. Loretz & J. Sanmartín (1995), The *Cuneiform Alphabetic Texts from Ugarit, Ras Ibn Hani and Other Places: (KTU: Second, enlarged edition)* (2nd ed. of Keilalphabetischen Texte aus Ugarit, 1976. ed., Abhandlungen zur Literatur Alt-Syrien Palästinas; Bd)
LHBOTS	Library of Hebrew Bible/Old Testament Studies
LS	Long & Sedley (Long, AA. & Sedley, D.N. (1987) *The Hellenistic Philosophers*. Cambridge: Cambridge University Press)
LSJ	Liddell, Scott & Jones, *A Greek-English Lexicon*
LXX	Septuagint
MT	Masoretic Text

Abbreviations

NIV	New International Version
NJB	New Jerusalem Bible
NRSV	New Revised Standard Version
SBL	Society of Biblical Literature
SBLABib	Society of Biblical Literature Academia Biblica
SOTSMS	Society for Old Testament Studies Monograph Series
SVF	*Stoicorum veterum fragmenta* (ed. von Arnim, H.F. 1903–1905 (reprint in 1964 by Teubner))
TAPA	*Transactions of the American Philological Association*
Tg.	Targum(s)
v.	Verse
VT	*Vetus Testamentum*
VTSup	Vetus Testamentum Supplements
Vulg.	Vulgate
vv.	Verses
WBC	Word Biblical Commentary
ZAW	*Zeitschrift für die alttestamentliche Wissenschaft*

Classical Greek and Roman works

Ag.	Agamemnon		*Metaph.*	Metaphysics
Ant.	Antigone		*Mete.*	Meteorology
Cho.	Libation Bearers (Choephori)		*Nat. D.*	On the Nature of the Gods
Criti.	Critias		*Op.*	Works and Days
DRN	On the Nature of the Universe		*OT*	Oedipus the King
Ecl.	Eclogues		*Part. an.*	The Parts of Animals
Enn.	Enneads		*Pers.*	Persians
Ep.	Epistles		*Ph.*	Physics
Epict. diss.	Discourses of Epictetus		*Phdr.*	Phaedrus
Eth. Nic.	Nicomachean Ethics		*Plt.*	Statesman (Politicus)
Eum.	Eumenides		*Pol.*	Politics
Fast.	Fasti		*QNat.*	Natural Questions
HN.	Natural History		*Resp.*	Republic
Il.	Iliad		*Sept.*	Seven Against Thebes
In Tim.	Commentary on Plato's Timaeus		*Sib. Or.*	Sibylline Oracles
Leg.	Laws		*Ti.*	Timaeus
Med.	Meditations			
Menex.	Menexenus			

Abbreviations

Bible books and other Second Temple works cited in this volume

Hebrew Bible/Old Testament

Gen.	Genesis
Exod.	Exodus
Num.	Numbers
Lev.	Leviticus
Deut.	Deuteronomy
Josh.	Joshua
Judg.	Judges
2 Sam.	2 Samuel
1 Kgs	1 Kings
2 Kgs	2 Kings
2 Chron.	2 Chronicles
Neh.	Nehemiah
Job	Job
Ps. (pl. Pss)	Psalm(s)
Prov.	Proverbs
Eccl.	Ecclesiastes
Song	Song of Songs
Isa.	Isaiah
Jer.	Jeremiah
Lam.	Lamentations
Ezek.	Ezekiel
Dan.	Daniel
Hos.	Hosea
Joel	Joel
Amos	Amos
Jon.	Jonah
Mic.	Micah
Nah.	Nahum
Hab.	Habakkuk
Zeph.	Zephaniah
Zech.	Zechariah

Apocrypha

2 Esd.	2 Esdras/4 Ezra
Wisd.	Wisdom of Solomon
Sir.	Ben Sira (Ecclesiasticus)

New Testament

Mt.	Matthew
Mk	Mark
Lk.	Luke
Jn	John
Rom.	Romans
1 Cor.	1 Corinthians
Eph.	Ephesians
Col.	Colossians
Rev.	Revelation

Other Second Temple Works

Jub.	Jubilees

Dead Sea Scrolls

1QpHab	Habakkuk Pesher
4Q416	Sapiential Work A^b/The Children of Salvation

CONTRIBUTORS

Professor Robin Attfield
Professor Emeritus of Philosophy, Cardiff University, UK

Robin Attfield is currently researching for a commissioned essay on panentheism, and also for a commissioned book to be entitled *Environmental Thought: A Short History*. His latest book was *Wonder Value and God* (Routledge, 2017), and his short book *Environmental Ethics: A Very Short Introduction* is due to be published by Oxford University Press in December 2018.

Dr Emmanuela Bakola
Associate Professor in Ancient Greek Language and Literature, University of Warwick, UK

Emmanuela Bakola has published widely on Greek tragedy, comedy and satyr drama, including the monograph *Cratinus and the Art of Comedy* (Oxford, 2010). She is currently working on a new book on the *Oresteia,* entitled *The Erinyes and the Wealth of the Earth: Cosmos, Nature and Resources in Aeschylean Theatre*.

Dr Ailsa Hunt
Lecturer in Ancient History, University of Birmingham, UK

Ailsa Hunt publishes on Roman religion and the history of its scholarship. She has strong research interests in the intersection of theological thinking and thinking about the environment in the Roman world, and has been working to revitalize interest in the natural world among scholars of Roman religion. *Reviving Roman Religion: Sacred Trees in the Roman World* was published in 2016 by Cambridge University Press.

Professor Christoph Jedan
Professor of Ethics and Philosophy of Religion in the Faculty of Theology and Religious Studies, University of Groningen, Netherlands

Christoph Jedan's research interests include Ancient Greek and Roman philosophy, the intersections of religion and philosophy today and the history and continuing relevance of consolation for death and loss. He has edited and co-edited half a dozen books and special journal issues, including *Stoic Virtues: Chrysippus and the Religious Character of Stoic Ethics* (Continuum, 2009) and *Consolationscapes in the Face of Loss: Grief and Consolation in Space and Time* (Routledge, 2018).

Contributors

Professor Melissa Lane
Class of 1943 Professor of Politics and Director of the University Center for Human Values, Princeton University, United States

Melissa Lane's books include *Method and Politics in Plato's Statesman* (Cambridge University Press, 1998); *Plato's Progeny* (Duckworth, 2001); *Eco-Republic* (Peter Lang, 2011/Princeton University Press, 2012); and *Greek and Roman Political Ideas* (2014; revised edition published as *The Birth of Politics*, Princeton University Press, 2015).

Dr Hilary Marlow
Fellow and Graduate Tutor, Girton College and Affiliated Lecturer, Faculty of Divinity, University of Cambridge, UK

Hilary Marlow's current research is on the relationship between God, people, and land/wild nature in the Old Testament/Hebrew Bible. Her publications include *Biblical Prophets and Contemporary Environmental Ethics: Re-Reading Amos, Hosea and First Isaiah* (Oxford University Press, 2009) and numerous articles on eco-critical readings of biblical texts.

Professor David Sedley
Emeritus Laurence Professor of Ancient Philosophy, University of Cambridge, and Fellow of Christ's College, UK

David Sedley's recent work includes the editing of philosophical papyri, Xenocrates' Platonism, and ancient theology and teleology. His most recent monograph is *Creationism and Its Critics in Antiquity* (Berkeley, 2007).

Dr Helen Van Noorden
Senior Lecturer and Fellow in Classics at Girton College, Cambridge, UK

Helen Van Noorden's research focuses on discourses rooted in archaic Greek epic, especially ancient didactic and eschatological literature of all kinds. She is the author of *Playing Hesiod: the 'Myth of the Races' in Classical Antiquity* (Cambridge University Press, 2014) and of articles on ancient receptions of Hesiod and on aspects of the Sibylline Oracles.

Dr Rebecca Watson
Research Associate at the Faraday Institute for Science and Religion, Cambridge and Dean of Studies at St Hild College, Yorkshire, UK

Rebecca Watson is the author, together with Meric Srokosz, of *Blue Planet, Blue God: The Bible and the Sea* (SCM Press, 2017). Her primary research focus is on the sea in the Hebrew Bible, understood both from ecological and history of religions perspectives, and on the Psalms.

Contributors

Dr Jula Wildberger

Professor of Classics and Comparative Literature at the American University of Paris, France

Jula has published widely on Hellenistic and Roman Imperial philosophy and literature. Recent publications include *The Stoics and the State: Theory – Practice – Context* (Nomos, 2018) and *Seneca philosophus*, edited together with Marcia L. Colish (De Gruyter, 2014).

INTRODUCTION
Ailsa Hunt and Hilary Marlow

Ecology and theology

Concern about the well-being of Planet Earth and the human impact on it is increasing, and new challenges are coming to light on, it seems, a weekly basis. As we write this Introduction, the problem of plastics and single-use packaging has come to the fore, with concerns raised by scientists that a vast expanse of non-biodegradable waste is contaminating our oceans, poisoning wildlife and, ultimately, entering the food chain. This is just one recent example of an ever-worsening degradation of the physical environment at the hands of human beings that threatens our own futures as well as that of countless other species.

Environmental philosophers and scientists both maintain that we have left the Holocene era (the period of geologic time that began about 11,700 years ago) and are now entering a new epoch, the Anthropocene, 'the age of humanity, when our actions are transforming the world' (Purdy 2015: 16). The term Anthropocene was first proposed by Nobel-prize-winning chemist Paul Crutzen in 2002, as a way of representing the complex array of changes brought about by human development that are having a permanent and often detrimental effect on the planet (Crutzen 2002).[1] Crutzen cites numerous factors, including the rapid expansion of human population, the exploitation of earth's natural resources, increase in concentrations of the 'greenhouse gases' carbon dioxide and methane in the atmosphere, and over-fishing and unsustainable land use. Jedediah Purdy regards the notion of the Anthropocene as primarily a cultural rather than geological idea, with political and ethical potential, and suggests that it is 'the starting place for a new politics of nature, a politics more encompassing and imaginative than what we have come to know as environmentalism' (Purdy 2015: 17).

Alongside calls for political and economic change, the role of religion in moderating and transforming community attitudes towards nature is increasingly recognized, especially in non-Western societies, where the majority of the population is religious.[2] Environmental scientists are noticing a geographical correlation between biodiversity 'hotspots' around the world and areas with high religious adherence (Bhagwat et al. 2011). And religious leaders are recognizing a responsibility to mobilize these communities to environmental action. In his 2015 Encyclical, Pope Francis addressed the loss of connection between people and the natural world, blaming exploitative industrial and technological practices and economic greed for the worsening environmental situation and issuing a resounding ethical call, to people of all faiths and none, to 'heed the cry of the poor and the cry of the earth' (Francis 2015: §50). Of course, Christian leaders are not alone in seeing the need to take a stance on environmental devastation. In fact, as Roger Gottlieb argues, all world religions have now confronted their attitudes towards nature and their potential complicity in the crisis, to the degree that he can claim: 'world religion has entered into an "ecological phase"' (Gottlieb 2006: 6). In this volume, we argue that, in ancient religions too, ecology and theology intersected. Hence this is not, as Gottlieb implies, only a phenomenon that has arisen in the last fifty years or so. Moreover, it is a major claim of this volume that we should not only be looking to contemporary world religions when facing the enormity of our environmental predicament, but that ancient religious thinking also has been, and should be, informing productive and creative responses to the challenges we face.

Why ancient religions?

This claim may come as a surprise: for awareness of environmental damage as the problem of *our* times can blind us to the fact that inhabitants of the ancient world (c. 800 BCE–400 CE) were also acutely aware that they existed as part of an ecological system. Although they did not use the language of ecology or environment (terms we shall discuss later in this Introduction), they were in many ways much closer to the natural world than our own heavily industrialized and technological era, and more aware of both the benefits and the dangers of nature. They understood, in a way that we have forgotten, where their food came from and recognized the importance of the agrarian cycles of planting and harvesting. They also lived more precariously than we do, at least in the West, and were at the mercy of drought and floods. Trying to make sense of the material world in which they found themselves easily led to asking bigger questions about human identity and roles within the world: consequently, when ancient thinkers considered human relationships with the natural environment, it usually did not take long for questions about God/the gods to enter the picture. Indeed, as we shall see in this volume, many ancient authors show themselves to have internalized a kind of conceptual triangle of connections between nature, humans and God/the gods.

The concept of the 'ecological triangle' as a lens through which to read ancient texts is one that has been developed in recent studies of the Hebrew Bible (e.g. Marlow 2009; Wright 2004); it can also be fruitfully applied to ancient thinking about the relationship between humans and the physical earth in other texts. The premise of such ecological hermeneutics is that in the ancient world, as in modern society, people cannot be viewed in isolation from the environments they inhabit, either physically or morally.[3] As used by biblical scholars, the ecological triangle seeks to identify ways in which particular biblical texts exhibit interrelationship – between God, humanity and the physical earth. This approach differs from most traditional and contemporary readings of the Hebrew Bible that prioritize the importance of human beings in the text and focus exclusively on the relationship between God and humanity. If 'God' is swapped for 'gods', this conceptual triangle offers a productive approach to the cultures of Greece and Rome also. By using this model to interrogate authors and their texts, hitherto neglected themes and motifs can be uncovered. The interrelational approach asks how the non-human world is understood and what assumptions are made about the relationship of God/the gods to the world and how they act within it. It also explores what effect the actions and choices of human beings have on the non-human world.

The idea of this internalized conceptual triangle between nature, humans and God/the gods is another way of framing the claim illustrated in this volume that ecology and theology are deeply intertwined in the ancient world. To think about human relationships within and with the world was always, also, to think about human relationships with the divine. Yet the intersection of ecology and theology within ancient cultures has been little recognized and so the first major aim of this volume is to paint a cross-disciplinary picture of ancient religious thinking about the natural world and to celebrate its richness. By putting theological ideas about the environment under the spotlight in this way, this volume takes a major and innovative step forwards in contributing to our knowledge of the ancient world. At the same time, the volume does not lose sight of our current ecological crisis. Rather it asks a central question: what impact, if any, should ancient ideas about the environment have on our own ecological thinking?[4] Moreover, the two aspects of this volume overlap and work in tandem. For the volume aims, in a mutually reinforcing process, to shape both our knowledge of ancient thinking and that of those who are writing the theology, philosophy and ethics of the twenty-first century.

Cross-disciplinary dialogue

The new status and relatively small size of discourse on environmental matters in a range of academic disciplines (those embraced in this volume being Biblical Studies, Classics, Philosophy and Theology) mean that this is an area ripe for further development and vitally in need of cross-disciplinary interaction. This volume sets out both to develop the field and to highlight fruitful areas for further study and cross-disciplinary engagement. It comprises the proceedings of a conference organized by the editors and held in Cambridge in 2014. Entitled 'Greening the Gods: Ecology and Theology in the Ancient World', the conference drew scholars working in the fields of contemporary ecotheology and philosophy, as well as scholars from a variety of disciplines with interests in ecological interpretations of ancient texts. It provided a feast of new ideas and approaches and the opportunity to engage critically with texts and authors, ancient and modern. This volume brings together essays in a number of disciplines from contributors to the conference, to give a fuller, more rounded view of what ancient 'ecotheology' might look like. In so doing it enables us to learn from other disciplinary approaches and techniques, and to identify both lacunae in understanding as well as areas of shared knowledge. The tendency for much academic study to take place within its own bubble, insulated from other scholarly disciplines, can mean that developments in other academic disciplines go unnoticed. This is very noticeable in the broad area of textual studies, with smaller disciplines such as Biblical Studies and Classics in some ways lagging behind Literary Studies when it comes to conceptual and methodological approaches. For example, the subject area of 'ecocriticism' was well underway in Literary Studies by the mid-1990s, and has now developed into a vibrant and burgeoning field, yet we rarely encounter acknowledgement of its concerns or methods by biblical scholars, classicists or ecotheologians.

The choice to adopt a cross-disciplinary approach for the conference, rather than one rooted in a particular field of study, came out of the editors' own research interests and experience. Both Ailsa Hunt (a classicist) and Hilary Marlow (a biblical scholar) have worked extensively on environmental themes in the ancient world, whilst reading very different texts. In conversation together we became aware that the ecological triangle of relationships outlined above was a shared feature of many of the texts we studied. This is unsurprising since biblical texts and Greco-Roman ones are from (broadly) similar time periods and engage with similar political and social issues. Both biblical and classical texts demonstrate a concern with nature and natural resources; whether or not they are explicitly religious texts, all operate within a religious socio-cultural context. Despite the relative lack of comparative studies between biblical and classical texts, there is potential overlap in both the sources of, and background to, such ancient texts. The groups that produced and edited biblical or classical texts may well have been shaped by similar external factors or may even have operated within similar cultural milieu, allowing them the potential to influence each other.[5]

Key terms and concepts

There are of course a number of difficulties inherent in a cross-disciplinary approach and it would be misleading to suggest that this is without its challenges. Some of these are at the level of terminology, chiefly regarding its cultural appropriation and the spread of (popularized) terms across disciplinary boundaries. In this section we will steer through some of the minefields surrounding the language used, and discuss the origin and meaning of key ideas, as well as their contemporary usage.

In the last decade or so, the nouns 'ecology' and 'environment' (or their respective adjectives, 'ecological' and 'environmental') have become familiar terms in the English-speaking world. The generic prefix 'eco' has been added to existing words to denote new forms of political and social discourse that address current concerns about the state of the planet, such as eco-capitalism, eco-feminism and eco-modernism. 'Eco' has come to denote anything purporting to have environmentally sustainable credentials (eco-friendly, eco-label) and has also been adopted by numerous businesses, selling anything from cleaning products (Ecover) to renewable energy (Ecotricity). In the commercial world it is now a major selling point to be able to apply the 'eco' tag to your product (with or without genuine justification). Likewise 'the environment' has become shorthand for anything to do with the physical world in which we live, and terms such as 'environmental crisis' and 'environmental damage' are used interchangeably to highlight the effect human beings are having on the planet.

These expressions, 'ecology' and 'the environment', are often used as synonyms, in both public discourse and scholarly writing (and indeed by numerous contributors to this volume). Some major academic works on ecological themes make no attempt to define the two terms and sometimes even appear not to distinguish between them, presumably for the sake of linguistic variety. For example, in his Introduction to *The Oxford Handbook of Religion and Ecology* Roger Gottlieb refers, within the same short section, to 'ecological religion' and 'religious environmentalism', essentially describing the same thing (Gottlieb 2006: 8–10). Despite this elision, there are subtle but important distinctions between the terms, especially as originally construed.

It was a German zoologist, Ernst Haeckel, who first coined the noun 'ecology' (Ökologie) in 1866 to describe the study of relationships between organisms and their immediate surroundings (Cittadino 2006: 74). Haeckel's choice was deliberate and based on the Greek term *oikos* 'household' (from which, in a separate move, the English term 'economics' derives). Initially reserved for scientific study of nature, it was not until the latter half of the twentieth century that ecology took on a new sense 'as a philosophy, a general worldview or a holistic vision' (Cittadino 2006: 93), and that a connection or even overlap with notions of 'the environment' developed.

The term 'environment' derives etymologically from the French *environer* 'to surround'. Hence the original sense of the English word, dating, according to the *Oxford English Dictionary*, from 1603, is the action or state of being surrounded or encircled. The use of the term in its most common and prevalent sense to denote 'the conditions or influences under which any person or thing lives or develops' was first coined by Thomas Carlyle in 1828 as a translation of the German 'Umgebung' in a literary essay on Goethe's work (Jessop 2012). In the context of rapid industrialization in the Scottish Lowlands in which Carlyle wrote, together with the shifting of cultural and literary boundaries, this coinage represented, according to Ralph Jessop, 'an act involving cultural internationalism, expansiveness, indefiniteness, and a mixture of literary, philosophical, and social critique', and so presaged the complexity of contemporary environmental concerns (Jessop 2012: 709).

Modern environmentalism with its strong moral stance and ethical implications draws on the work of nature writers such as Henry David Thoreau and John Muir, and scientists such as Rachel Carson, who each 'sought to defend vulnerable nature from self-seeking, thoughtless modern industrial society and to spread appreciation for nature's beauty and wonder' (Stoll 2006: 66). Muir and Carson were certainly influenced in this by their strong Presbyterian roots, which sought to protect the 'Eden' of the American Mid-West from exploitation and commercialism (Stoll 2006: 66). At the forefront of the development of the field of environmental ethics is Aldo Leopold, whose seminal and popular work *A Sand Country Almanac* developed the notion of a community of nature into an ethical mandate (Leopold 1987). Leopold argued for a 'land ethic' that accorded moral standing to non-human entities, including soil, water and plants as well as animals. His Golden Rule of ethics states: 'A thing is right when it tends to preserve the integrity, stability, and beauty of the biotic community. It is wrong when it tends otherwise' (Leopold 1987: 224–5). Leopold is regarded as 'the father or founding genius of recent environmental ethics' (Callicott 1989: 15), and provided the catalyst for subsequent generations of environmental philosophers as well as ethicists.

Despite the blurring of the distinctions between the terms 'ecology' and 'environment', it remains the case that by and large ecology describes the systems that operate, while an environment is the product of those systems. Notwithstanding this distinction, as we have seen, the prefix 'eco' has proved a useful shorthand for anything connected to environmental concerns, and functions in much the same way as the adjective 'green' in a variety of contexts. One particular example is the term 'ecotheology' already used several times in this Introduction of both modern and ancient authors. First adopted in the late twentieth century in Christian circles, ecotheology is concerned with the intersection of religion with nature, often, though not always, with a particular focus on environmental concerns. The modern ecotheologians discussed in this volume (e.g. by Hunt) are certainly intentionally responding to contemporary concerns; the ancient writers are doing 'ecotheology' insofar as they are exploring the connections between human beings, nature and God/the gods.

This leads to another fundamental issue arising from use of the terms ecology and environment: the relationship between the ancient world and contemporary understandings of environmental issues. The authors of ancient biblical and classical texts could not have envisaged the scale of human impact on the land in the twenty-first century (Marlow 2009: 8). Relating these modern terms and developments back to the ancient world may result in anachronism, a danger against which scholars such as Cyril Rodd have warned in no uncertain terms (Rodd 2001: 249). But perhaps this is to miss the point. Even though the scale of environmental damage we are experiencing today is unprecedented, and the terms we use are less than 200 years old, it does not follow that ancient Israelites or Greeks or Romans were oblivious to their surroundings, as we have already noted. Rather many ancient texts display awareness of a complex relationship between human beings and the natural world. To avoid charges of anachronism, however, some might wish to adopt an alternative term to 'environment', with all its cultural and temporal ties, especially, though not exclusively, when speaking of ancient contexts. One attractive option is 'landscape', which environmental philosopher Holmes Rollston III uses to distinguish between human and non-human spheres. Rollston denotes landscape as 'the shape of nature, modified by culture from some locus', adding that 'in that sense landscape is local, located' (Rolston III 1995: 379, see also Hailwood 2004: 24–5). Whatever terms we use, it remains the case that each of the ancient authors discussed in this volume lived and wrote from within a particular landscape, and that local intersection of nature and culture shaped their thinking and writing about relationships between humans and their natural surroundings (environment, if you like) – or in other words, ecological relationships.

Hermeneutical questions

Reading ancient texts in the light of contemporary issues always throws up hermeneutical questions and challenges (how should a feminist read Ovid?; or how should a liberation theologian read Paul?). Using ancient ecotheology to think through our environmental crisis is no exception to this rule. The pitfalls are obvious. Cicero of course had no opinions on the problem of single-use plastics, because it was not a problem. Whilst ancient cultures are by no means totally innocent of environmental change and damage (and occasionally show awareness of this, e.g. Theophrastus, who in *On the Causes of Plants* 5.14.2–3 acknowledges that the draining of a lake in Thessaly has affected the local climate), they did not cause anything like the scale of environmental damage which modern technology and attitudes have created.[6] There is thus a real danger of forcing the ancient evidence to make it relevant and an anachronistic application of terms such as 'green'.

Since there was no ancient environmental crisis on which to reflect, we can hardly expect ancient authors to provide us with ready-made theories as to how to respond to our own. But this need not be a negative. For what we need in responding to our own environmental challenges is not another policy on sustainable fishing or a new idea for renewable energy production, but rather a global change in individuals' attitudes to the environment. As early as 1998, the year in which the Kyoto Protocol was adopted, such a call for moral transformation to enable 'responsible stewardship of the earth' was made by Maurice Strong, former UN Under-Secretary-General: 'In the final analysis, our economic and social behavior is rooted in our deepest moral and spiritual motivations' (Strong 1993). It seems that scientific data and reasoning is not enough to change our mindsets, as Professor Gus Speth, the former Dean of Yale School of Forestry, summed up:

> I used to think that top global environmental problems were biodiversity loss, ecosystem collapse, and climate change. I thought that with 30 years of good science we could address these problems, but I was wrong. The top environmental problems are selfishness, greed, and apathy, and to deal with these we need a spiritual and cultural transformation. And we scientists don't know how to do that. (Wilson 2008, see also Speth 2008: 199–216)[7]

In other words, we do not need better science, or new and improved versions of what we already have, but rather a radical shift in our thinking, an outside-of-the-box kind of thinking.

Ancient ways of thinking about the world – with their 'otherness' as well as their perceived timelessness and their enduring appeal – could be just the kind of catalyst needed to prompt modern audiences to such a 'spiritual and cultural transformation' and change in mindset. This, for example, is what Melissa Lane found when studying Plato's *Republic* in the light of debates over climate change (Lane 2011). Like Strong and Speth, she too argues that our apathy in responding to environmental challenges is 'in large part a failure of the collective imagination' (2011: 8), and in her research she encountered numerous individuals from all walks of life who (interestingly) spoke of the need for transformation as the 'need [for] a new mythology' (2011: 9). Lane proposes Plato as 'a possible source for stimulating reform' (2011: 12), thanks to his 'transformative images' (2011: 24), which have 'haunted readers over the centuries' (2011: 22), and which model 'what transformative social change would look like and require' (2011: 22).[8] The fact that ancient authors such as Plato were not themselves facing environmental challenges, as we are, means that they offer the modern reader not policies or advice, but ethical principles and frameworks that can then be applied to our own, ever-changing, situation. And because these ancient ideas are products of cultures temporally, and often geographically, very distant from our own, they come coloured with a sense of global and enduring relevance that may heighten their attractiveness.

Further questions about the relevance of these ancient texts are also thrown up by the cross-disciplinary nature of this volume, and the unsettling exposure to different disciplines' habits and methodological practices that this entails. Differences centre on understandings of the nature of the ancient texts, the lasting value of those texts outside of their original context and how those texts are to be read in relationship to one another.

The Bible represents a small and distinct corpus, regarded as a canon or 'library' of sacred texts from approximately the second century BCE (for the Hebrew Bible), and the third century CE (for the New Testament). These texts by very definition held theological and ethical meanings that were intended to be understood and acted upon by their earliest hearers/readers, and by subsequent generations of Jews and Christians. And because these texts have for centuries been read as a group and have impacted societies as a group, they are read in dialogue with each other, despite very different literary genres and, potentially, historical and temporal contexts. This means that a biblical scholar can legitimately bring a passage from one part of the canon into dialogue with a verse from another part. This is not to say that all biblical scholars are operating from a confessional perspective – many consider themselves first and foremost to be ancient historians and textual exegetes. However, it does impact on the methods adopted, and in particular the relative ease with which biblical scholars study biblical texts thematically and intertextually (see, e.g., Dell Kynes 2012). Moreover, within confessional Jewish and Christian theology, biblical texts are *expected* to be of contemporary relevance and application, and research undertaken from a confessional perspective often aims to show how a contextualized reading of these texts can make them speak to Jews and Christians today.

The corpus of Greek and Latin texts is not only much larger than the biblical corpus, but the fact that these texts are not considered scripture, or part of a formal canon, means they are not treated as 'members of the same family', as are biblical texts. A classicist, despite the weighty emphasis on intertextuality in classical scholarship,[9] would work to justify reading Aeschylus and Homer as 'parallel voices' in the same way as a biblical scholar might read, say, Isaiah and Psalms. They may also be unfamiliar with the kind of thematic approach to ancient texts such as Watson adopts in this volume. In addition, a classicist would probably need to work harder than a biblical scholar, and certainly than a Jewish or Christian theologian, to argue that ancient texts are of direct relevance to modern ethical concerns.

Meanwhile, philosophers who turn to classical texts work from an awareness that ancient philosophy was often developed in conversation/competition with previous philosophical positions; consequently it makes eminent sense to continue to read these texts in dialogue, and to use insights drawn from different texts in a collaborative way when

Introduction

constructing new interpretations of ancient philosophical theories. Whilst the corpus of ancient philosophical texts is not considered 'bound' in the same way as the texts of the biblical canon, philosophers reading ancient texts will often view the texts on which they focus as texts that work in tandem.[10] Moreover, these philosophers approach their texts recognizing that although ancient philosophical theories are the product of a particular context, those theories were never intended to be of relevance only in one particular place and time: rather they were conceived as modes of understanding the world with universal validity, and as such they have ample potential to speak to current situations.

This volume thus brings together diverse ways of thinking about the relevance of ancient texts for our environmental crisis, and the relationships between those texts, which will need to be treated with sensitivity. This will obviously aid dialogue, but will also be productive in other ways, opening up possibilities for the disciplines to learn from each other. Classicists and philosophers will likely always have to argue harder for the contemporary relevance of the texts they study to our environmental crisis than do biblical scholars and theologians speaking to pre-existing communities of readers of these ancient texts, who are keen to apply them to their own lived situation. But the Bible does not always obviously address environmental questions, so scholars who take a confessional approach to biblical texts could benefit from additional ways of arguing for ancient texts' environmental relevance. At the same time, biblical scholars and theologians may feel a far greater sense of urgency to set out what biblical texts say on environmental issues, than those scholars who work on classical texts. The latter could usefully harness some of that urgency and energy, which stems from a confessional engagement with the texts, to further debate about a crisis in which we are all implicated, regardless of our faith position.

Summary of chapters

The volume opens with Lane's parallel reading of ancient Hebrew, Greek and Roman texts, in which she emphasizes the place of politics in mediating between ecology and theology in the ancient world. Exploring the interrelated themes of land, nature and labour, the chapter ranges from the biblical promise of a land to the Israelites, to the Athenian myth of autochthony, to Greco-Roman images of the earth spontaneously producing food, to the account in Exodus of the plagues inflicted on the Egyptians. Unearthing the submerged role of politics in these ancient accounts of human interaction with the land, Lane not only casts new light on ancient ways of thinking about the intersection of ecology and theology, but also shows us that human political influence over the earth is nothing new. A reminder of that power, Lane argues, should give us pause for thought as we now seek to channel it into more ecologically conscientious forms.

Next Van Noorden's chapter puts the *Sibylline Oracles* in the spotlight. This idiosyncratic and challenging text blends archaic Greek epic, Hellenistic Jewish and early Christian traditions, whilst purporting to be prophecies of the Sibyl, who frequently depicts and prophesizes environmental disaster, often of apocalyptic proportions. Van Noorden argues that a central preoccupation of this Sybilline voice is the interface between the divine will, humanity and the environment, and that foregrounding environmental concerns offers a key to reading her prophetic idiom. Throughout the chapter she draws comparison with Virgil's *Eclogues*, another ecologically significant collection of texts, in order to enhance the clarity with which we see the Sibyl's own ecological thinking. Van Noorden explores this through four areas: one, the presentation of the land; two, the use of ecological metaphors to describe mythical and historical events; three, the overlap between natural disasters and acts of god. Fourthly, Van Noorden turns to ecocriticism, borrowing a model of reading texts from another discipline (namely literary criticism) to argue that, when thought of as ancient ecocritical literature, thematic and formal aspects of the *Sibylline Oracles* are mutually reinforcing. Ecocriticism encourages its readers to activism in the here and now, and Van Noorden shows that classical texts are also ripe for such an approach.

From here we turn to a cluster of philosophical chapters. Sedley explores the common idea in Greek philosophy that the world is a deity and a self-sufficient being. With particular focus on Empedocles' Sphairos and the depiction of god in Plato's *Timaeus*, he unpacks the Greek idea of cosmic self-sufficiency and draws out implications for our environmental crisis via affinities with two modern ideas: the Gaia hypothesis and the superorganism. Particular attention is given to the *Timaeus*' picture of the world god: spherical, fully self-sufficient and of everlasting durability, thanks to its ability to internally recycle all waste. Sedley questions how this picture can speak to the pressingly urgent issues concerning the world's self-sufficiency that we face today.

Next, Jedan and Wildberger offer a complementary and contrasting pair of chapters on the usefulness of ancient Stoic thinking about the world – which for the Stoics is also god – when constructing responses to our environmental crisis (a blossoming area of academic debate). Both conclude, in different ways and to differing degrees, that Stoic virtue ethics can productively inform modern environmental ethics. The crux of Jedan's argument is that in existing debate about Stoicism and environmental ethics one element has been largely ignored, namely Stoic theology. This Jedan puts to rights, arguing that there is no Stoic theology in the singular: what is often referred to as Stoic theology is in fact the interaction of three different theologies – pantheism, polytheism and henotheism – which are not fully compatible. Jedan goes on to show how the balancing of these different theologies led to Stoics adopting a moderate anthropocentrism. Jedan then investigates what a theologically focused understanding of ancient Stoicism can offer to modern environmental ethics. There are advantages – the focus on individual virtue offers an exellent antidote to the fear of individual powerlessness against the environmental crisis – but also disadvantages, such as the fact that Stoics did not view the world-god as something vulnerable. Jedan concludes that Stoicism offers us a 'light green' model of environmental thinking.

Wildberger takes as her starting point the central Stoic concept of *to kalon*, the beauty that Stoics ascribe to the cosmos, and argues in detail that this beauty should be understood as an expression of the world-god's sociability. She then argues that this interpretation of *to kalon* could help to ground modern environmental virtue ethics. The emphasis on sociability, Wildberger argues, helps to overcome the dilemma between individualistic and holistic approaches in modern environmental ethics. For individual interest, in her interpretation of ancient Stoicism, in fact resides in being instrumental and working as part of a whole; humans flourish in loving and caring for what is other than themselves. Interpreting ancient Stoicism in this way also, she argues, provides the best possible compromise between clashing interests of pre-existing entities, which is central to many decisions humans make about environmental action. Finally, Wildberger also reflects on the gap between ancient thinking about the environment and our own, considering how this Stoic idea can be translated to a context which does not presuppose theistic beliefs.

Rounding off this philosophical cluster, Attfield explores ancient philosophical and religious roots of the modern responses to environmental challenges, covering a huge array of authors from Aristotle through to Origen. Two themes come to the fore as areas on which all these authors have something to contribute: anthropocentrism and the idea of nature's intrinsic value; and human responsibilities for nature. Attfield explains how some of these ancient authors have been interpreted and used in the last decades of environmental debate, assessing the validity of these uses: Callicott's use of Plato's *Timaeus* and Passmore's use of ancient Stoicism, for example, are up for consideration. Attfield paints a picture of the breadth of ancient influence on modern environmental thinking, arguing that whilst the relevance of Darwinism, or the science of ecology, must not be downplayed, ancient influence is omnipresent. Even global agreements such as the Millennium Development Goals (2000) and the Sustainable Development Goals (2015) bear the marks of the ancient world.

The next three chapters focus our attention on the sea and earth. First Watson explores conceptions of the sea in the Hebrew Bible. Watson argues that these are shaped far more by the biblical idea of humans being creatures within creation, than the better-known notion of humans as stewards of that creation. Covering texts including the Psalms, Ezekiel, Jonah, 2 Esdras and Job, Watson identifies and analyses four themes: human fear and vulnerability in the face of the sea; the presentation of attempts to achieve mastery over the sea as a hubristic infringement of human limits against the divine; awe and appreciation of God's care for sea creatures and of his non-anthropocentric focus; and awareness that the sea is not a limitless resource. Watson delves into these themes in order to question whether they can contribute to and challenge contemporary responses to the environment. She concludes that the idea of being creatures within creation provides a model for environmental engagement that, although very uncomfortable to a modern audience, is increasingly recognized as necessary for ecological flourishing.

Next Bakola turns our attention to the earth, especially as conceived in Aeschylus' trilogy the *Oresteia*. Bakola argues that the category of chthonic deities (from the Greek *chthon*, meaning earth) is an obvious place to look for ecological awareness in the ancient world. While previous scholarship has focused on the underworld connotations of the term chthonic, Bakola argues for a more ecologically sensitive understanding of these deities, before turning to demonstrate what this brings to a reading of the *Oresteia*. Bakola focuses on the role of the Erinyes (chthonic deities) in the trilogy and in particular its final scene. Bakola explores how the earth informs understanding of the Erinyes and argues that Aechylus' presentation of them can be characterized as ecotheological: these are deities who avenge the waste and destruction of natural resources, as well as assuring natural growth. Bakola paints the *Oresteia* as a trilogy obsessed with a balance between humans, the earth and its wealth (and how this balance defines human relationships with the cosmic order), and one with a vital message for how we also manage that balance.

Marlow keeps our attention fixed on the suffering of the earth at human hands. Marlow explores the effect of warfare on the earth, first considering how this is portrayed in the prophetic texts of the Hebrew Bible, before bringing these texts into conversation with the poetry of the First World War. From the Hebrew prophets, in particular the book of Jeremiah, Marlow draws out persistent themes associated with depictions of warfare in these texts: the earth mourning or lamenting, the personification of nature, and the idea of cosmic disorder and de-creation. Through these themes Marlow analyses Hebrew thinking about the intersection of ecology and theology, especially the questions raised by the involvement of Israel's God YHWH in these earth-destroying wars. Marlow then asks how these themes might map onto depictions of the relationship between the earth and warfare in the poetry of the First World War. Throughout the chapter, Marlow reflects on the rationale for comparing these two distant collections, investigating lines of influence between them, the reasons behind their similarities and differences, and the degree to which their concerns are relevant to our own ecological situation.

The volume ends on a note of academic caution and cross-disciplinary introspection. Hunt's chapter calls attention to the power of a flawed idea about the ancient world to inform and shape responses to our environmental crisis. This is the idea that Greeks and Romans understood their natural environment to be animated by spirits. Hunt argues that Lynn White Jr., in his controversial and influential article *The Ancient Roots of our Ecologic Crisis* (1967), gave this idea (a) an official sounding label ('pagan animism'), (b) an ecological thrust – claiming that where Christianity had gone wrong, in environmental terms, was in its destruction of the world of pagan animism – and (c) life outside of his discipline. Next Hunt delves into the history of pagan animism, and the idea's continuing influence in recent classical scholarship, before going back to Greek and Roman evidence to argue that the idea is profoundly flawed. Hunt then charts the idea's cross-disciplinary impact, particularly within Christian ecotheology and biblical studies, and explores how undermining the validity of the concept of pagan animism might impact on environmental thinking within these disciplines. The take-home moral is that paying more rigorous attention to ancient ecotheological thinking can both challenge and rejuvenate modern environmental debate.

This too is the message of the whole volume. For ancient ecotheology is not only worthy of study in its own right, ignoring which is to undermine the richness of ancient thinking about how humans, nature and the divine intersected. It is also the case that contemporary thinkers will be far poorer for not considering how ancient ideas have impacted, could impact and should impact on modern responses to the ecological situation that we, and all future generations, face. This volume aims to illustrate, in a rich variety of ways, how the dynamism of ancient ecotheology could act as a catalyst for a radical shift in modern environmental mindsets.

CHAPTER 1
ANCIENT IDEAS OF POLITICS: MEDIATING BETWEEN ECOLOGY AND THEOLOGY
Melissa Lane

'We are as gods', asserted Stewart Brand in 1968 in speaking of the human relation to the environment – and, he enjoined, we had better get good at it (Brand 1968). Yet the fact that distributed human agency has brought about what is now increasingly widely termed the Anthropocene, the proposal to define a geological era succeeding the Holocene in virtue of the newly 'central role of mankind in geology and ecology' (Crutzen and Stoermer 2000: 17), does not mean that a reverse intervention of a quasi-omnipotent and benevolent kind is as readily within our current agential capacities. The enormous human capacity for political neglect, treating carbon emissions as free externalities, is not so readily convertible into the converse capacity for political control.[1] That is to say that we may turn out to be more like Greek gods – rivalrous, quarrelsome, and sometimes ultimately hapless – than an idealized monotheistic one, in our inability to control or reverse the effects of our actions (as argued also in Lane 2016: 118).

Brand's assertion is often taken to herald a major novelty in humans' relation to the environment: previously we were not as gods, but now we are (though subsequent research debates the moment at which this happened, whether it should be indexed to the Industrial Revolution, or whether human interventions have had major effects on the biosphere for thousands of years before that, perhaps in bringing about the 'fifth extinction' long before currently threatening to bring about the sixth) (Kolbert 2014). Yet if this way of thinking is relatively new to the environmental movement, reflecting on the relationships between humans, god or gods and nature is much more deeply rooted in the history of human thought, flourishing in the Axial Age of synchronic great civilizations from about 800–200 BCE (Bellah 2005). Thinking about how human agency, in the absence or presence of political organization, can interact with nature, and how that agency compares to how one might imagine the divine, is engrained in the great texts of that era, and can perhaps provide some guidance as we ourselves confront these questions afresh. This article will explore three particular themes – labour, land and nature, especially as related to divine punishment and reward – in both ancient Greek

---

I am grateful to the editors for the honour of serving as the keynote speaker at the 2014 conference that gave rise to this volume: 'Greening the Gods: Ecology and Theology in the Ancient World' (University of Cambridge), and to all the participants therein; to participants in a 2017 faculty seminar on a later version of the paper at the Princeton Environmental Institute; and to Diana Lipton and Rob Socolow for more general advice and inspiration on the topic of this contribution.

texts and the Hebrew Bible, in order to see what both polytheistic and monotheistic approaches might have to offer in shaping our meditations on god, nature and human agency today.[2]

Of course, the role of antiquity, both pagan and monotheistic, in framing the relationship between humans, gods and nature, has long been a subject of discussion. More than fifty years ago, the distinguished medieval historian Lynn White, Jr, gave a lecture on 'The Historical Roots of Our Ecologic Crisis' that offered one of the earliest grand narratives – from the ancient world through the whole sweep of Western history – of the relationship between ecology and theology.[3] The lecture is generally remembered as a key intervention inaugurating the debate about the relationship between what White called 'Judeo-Christian teleology' and scientific and technological advancement, one that White concluded by urging a return to the alternative Christian path charted by St Francis of Assisi, whose entertaining of the idea of animal souls was lost in the dominant Christian narrative of God and man as transcending nature (White 1974: 23).

My interest here lies in an aspect of White's narrative that has received scant attention: the place of politics in his account. He wrote that the fusion of science and technology was primarily achieved by the advance of modern democracy in the nineteenth century, which 'by reducing social barriers' brought about the 'functional unity of brain and hand', and he summed up his case by stressing not so much theology or science but the underlying role of democracy: 'Our ecologic crisis is the product of an emerging, entirely novel, democratic culture' (White 1974: 19). Without focusing specifically on the modern democratic element, what interests me in this thesis is the centrality that it gives to politics, a topic that has been relatively neglected in the great debates about the relationship between theology, science and technology, and ecology in Western and non-Western histories. In this chapter I explore the place of politics – in historical reality, in political theory and in philosophical imagery – in mediating between ecology and theology in the ancient world, both in Greek and Roman ideas and in the Hebrew Bible.[4] Whereas in earlier work I have examined the relationship between politics and psychology as charted in Plato's *Republic* as a key element of social stability, and hence of social and ecological sustainability (Lane 2011/2012), here I explore the relationship between politics and nature more directly, arguing that to think about ecology and theology is also – perhaps unexpectedly – to think about politics. Politics is not just an adventitious reaction to environmental and ecological issues. It is in ancient thought an integral part of how such issues were understood to arise and to be addressed.

Politics in the human relationship to nature in – and before? – the Anthropocene

To sustain the value of this thesis for thinking about ecology today, we need to confront a major preliminary challenge: that the categorical gulf between the way that humans related to nature in earlier geological eras, and in what is now known as the 'Anthropocene', in which humans have become the dominant force shaping the climate of the earth, makes any lessons from that earlier relationship irrelevant. While not spelling out such an implication, Dipesh Chakrabarty has laid out the basis for such a challenge in his important 2009 article, 'The Climate of History: Four Theses'. There he argues as his first claim that while humans have always been 'biological agents', it is only recently that they have – 'historically and collectively' – become 'geological agents' in altering the chemistry of the atmosphere and so the basic conditions of climate, sea level and ice; he explains this geological agency as meaning that humanity now exerts 'a force on the same scale as that released at other times when there has been a mass extinction of species' – hence justifying the attribution of the new geological era of the Anthropocene (Chakrabarty 2009: 207).

There has no doubt been – if I may put it this way – a seismic shift in the nature of human agency in the Anthropocene. But Chakrabarty's contrast between past merely biological agency, and current and recent geological agency, obscures the role potentially played by politics in each time period, at whatever scale and nature of impact. We can illustrate this for example by considering in what way humans may have played a role in 'the disappearance of large mammalian fauna' in Roman North Africa (Shaw 1981: 385, for this quotation and the others in this paragraph). An earlier view was that human biological agency was not important at all – that such disappearance and other shifts in the ecological balance of the region in late antiquity simply and directly reflected determinist effects of changes in climate. Against that view was an intervention by Brent Shaw asserting a role precisely for human 'biological agency': he countered the argument for the simple 'effects of a deteriorating climate' in causing this disappearance by arguing that such effects 'cannot be sieved out neatly from the one major discernible cause, the hand of Man'.

Shaw had in mind changing hunting and agricultural practices. Those are themselves the products of human communities, mediated through linguistic and cultural inheritances and practices. Indeed, for the classical Greeks, a *politeia* (constitution) would have included the practices of a general way of life, from patterns of child-bearing and child-rearing to daily diet to exercise and education.[5] So I would want to go a step beyond his explicit argument, to suggest that such changing practices must themselves reflect political choices, possibilities, doings and allowings. This is not to reduce human agency to politics, but to insist that as political animals (as Aristotle would say), the political dimension of our agency is always to be considered – while asking how self-aware, collectively adequate to its ends and comprehensively engaging that dimension might in any given case be.

In other words, the practice of politics is likely already to have been part of what has shaped how humans have acted as what Chakrabarty calls biological agents. And for all of its novelty, the same is true of our geological agency in the present era. Both of these forms of agency are always at least potentially mediated by politics. And that role of politics in connecting humans, gods and nature is something that the ancients understood in their time, reflection on which can help us to better understand the new forms of interaction between humans and nature in ours. Ancient thinkers like Hesiod, Plato and the authors of the Hebrew Bible did much to trace the effect of humans on nature not simply as 'biological agents', whose effects on the natural world were the unthinking material results of biological processes, but rather as political agents who engaged with nature through the practices of politics. In their reflections on this point we find much that can help us both to appreciate the complexity and shape of ancient views themselves, and also to think for ourselves about the ways in which politics remains a critical factor in shaping the future of ecology in the Anthropocene. Whereas in ancient times the different spheres of human influence were spatially demarcated in relationship to the divinity or divinities apportioning them different plots of land, they are now temporally demarcated in geological time for the globe as a whole.

Spontaneous food supply and the absence of politics

Let us begin with some strikingly overlapping visions of an idyllic era in biblical and Greek texts, in which for the most part we find a spontaneously self-renewing sustainability of nature and an absence of politics. This will equip us, in the subsequent section of the chapter, to appreciate the role that agricultural labour plays in opening up a role for politics. Finally, we will turn to the condition of the land as the result of a complex interplay between divine rule and human agency, including political agency.

So to begin on the absence of politics and its relation to ecology and divinity: I will consider the two different narratives of the creation story in Genesis (in Hebrew, berēšît), and the three related but subtly different narratives of the Golden Age or Age of Kronos in Hesiod's *Works and Days*, and in Plato's *Statesman* and *Laws* (Lane 2017). In all of these we find a vision of a recurrent, perennial supply of food, paradigmatically in the form of seed-bearing plants that have self-renewing powers to provide food (in particular, fruits) while replenishing themselves for future growth. In other words, in the abundance of such plants, we find an intrinsically sustainable relationship between humans and nature – one that seems not to need politics to be maintained, though as we shall see, the danger that politics could undermine it is arguably foreshadowed in at least one of these accounts, in Plato's *Laws*. Once the golden age is left behind, however, humans will have to develop further arts of agriculture and artisanry as well as politics – and the future sustainability of their relationship to nature will be dependent on those arts not undermining or harming it, as they will have the power to do. These accounts may combine some dimension of cultural memory, of the stages of evolution of nature and culture (as with the contrast between pastoralism and agriculture in the story of Cain and Abel, Genesis 4), with fantasies of pure and spontaneous provision that could eschew labour altogether.

In the first biblical account, on day three of the creation of the world, God creates vegetation: 'seed-bearing plants of every kind, and trees of every kind bearing fruit with the seed in it' (Gen. 1.12). After human beings have been created on the sixth day, these seed-bearing plants and fruits are presented to them as 'yours for food' (1.29). In this first narrative of creation in Genesis, there is no mention of need for work or toil: the seed-bearing plants and trees constitute a sustainable and spontaneous source of food.[6]

Simultaneously, in the second creation account (Genesis 2) paradise is modelled as pertaining to a single conjugal pair, so excluding any role for collective politics.[7]

We find abundant fruit embodying the dream of food without toil in Hesiod and some works of Plato as well. In *Works and Days*, Hesiod describes the golden age which is under the kingship of Kronos thus (adding spontaneous pastoral animal reproduction to the spontaneity of fruit-bearing trees): 'spontaneously then/The earth bore rich, abundant fruit; and these contented men/Living in peace, enjoyed its works and all its many goods,/Abundantly supplied with sheep, beloved of the blessed gods' (116–20).[8] Notice that unlike in the Bible's initial portrayal of human life, here – among an entire race of humans, a plural collective as opposed to a single conjugal pair – war and political strife is conceivably possible. What rules it out, by implication, is the contentment that the golden race enjoys from the effortless sourcing of food (and sheep), hence leading to peace. In other words, a spontaneously sustainable food supply can make politics unnecessary.[9] By inference, the absence of such a spontaneously sustainable balance of food provision will make politics necessary, but in turn, also potentially inimical to that paradisical balance.

In Plato's *muthos* (story) in the *Statesman* (or *Politicus*), which directly references the Hesiodic 'golden age' under Kronos' kingship as one of its sources, the *daimonion* (divine assistant) figures who divide up living beings among them – each to rule one kind of living beings – ensure that there is 'no war or internal dissent' among (I take it) all living beings (*Plt.* 271e2).[10] The *daimonion* overseeing humans further ensures that humans may live 'without toil' (*Plt.* 271e4: *automatou*), with 'an abundance of fruit from trees and many other plants, which grew not through cultivation but because the earth sent them up of its own accord' (*Plt.* 272a2–5).[11] In this narrative, the absence of politics is not directly connected to the absence of toil for food afforded by the fruit trees, but rather to the more general condition of this era of direct tendance (expressed in the vocabulary of pastoral shepherding) by a divine being: 'given his tendance, they had no political constitutions' (*Plt.* 271e7–8). Still, the absence of toil for food, the self-renewing sustainability of the environment insofar as it meets human needs, and the absence of politics in this account go together.

Contrast the account of the age of Kronos given by the anonymous Athenian Visitor who is the main speaker in Plato's *Laws*. On the one hand, the Athenian Visitor (like his Eleatic counterpart in the *Statesman*) stresses that the humans of that age 'were provided with everything in abundance and without any effort on their part', glossing this as 'the wonderfully happy life people lived then' (*Leg.* 713c2–4). More important for our purposes, in the Athenian's account, the direct rule over humans by one or more *daimonion* figures is described as producing 'peace, respect for others, good laws, justice in full measure, and a state of happiness and harmony among the races of the world' (*Leg.* 713e1–3). Here, divine kingship and secondary-divine tutelage is compatible with politics of a certain kind – laws and justice – in addition to peace. In contrast to Hesiod who linked effortlessly abundant food directly (via contentment) to peace, the Athenian connects effortlessly abundant goods to a more active political vision, if one remaining under direct divine control. The moral is that a theology of divine rule can make room for a real role for politics, and in particular, for one that bears on the relation between humans and environment.

Agriculture and human impact

If Plato's *Laws* injects human politics into the age of Kronos, the Bible's second creation story in Genesis for its part injects human agriculture into paradise: the man is commanded 'to till ... and tend' the garden (2:15). Many interpreters have read this second creation story as a humbler one that is a better model for modern ecological principles: Daniel Hillel, for example, contrasts the 'anthropocentric' focus on humans as being given the 'right to dominate all other creatures' in the first creation story in Genesis 1, with their 'stewardship' role not as rulers but as 'custodians' in the second creation story in Genesis 2.[12] My focus here however is on the fact that, although the soil is cooperative and the labour not painful (as it will become with the expulsion from paradise), in this second biblical account work must be done to produce food, even in Eden.[13] This second biblical narrative reminds us that in fact, the dream of food without any toil is a fantasy. Even fruit trees need pruning if they are to flourish long term (though it is worth noting that they were the last kind of trees to be domesticated, in part, it seems, as their time to reproductive maturity is so long: Goldschmidt 2013). That is, even fruitarian food requires a cooperative venture between nature and human craft, all the more so insofar as human activities come to disrupt the health of the tree and plant self-reproduction with which divine creation originally endowed them. In its second creation story, the Bible recognizes this fact, acknowledging the partnership between humans and the natural world that sustaining food will require. Elsewhere the Bible also prohibits the destruction of fruit-bearing trees in the course of laying siege to a city in wartime (Deut. 20.19).

At the same time, the potential is opened here for humans to undermine ecological sustainability. If humans practise agriculture or mining in damaging ways, and if their political decisions either facilitate or fail to stop those damaging practices, then the human role can turn from productive to destructive. (One might say that Eden can continue just so long as humans take no initiative, but that once they do so, the environment becomes inherently vulnerable to human action from then on.) Indeed a brief but pregnant allusion to these sorts of possibilities is found in Plato's *Critias* (111a–d). There, as others interested in Plato and the environment have noticed,[14] Critias describes a process of soil erosion that has left the soil of modern Athens less productive – though still relatively more so than that of other lands – than was true of the land in the archaic period he is recounting. His main focus is on the natural causes of this erosion due to the high promontories down which soil ran off into the sea (*Criti.* 111b1–7). But he does mention that Athenian houses still sport rafters from trees cut down for wood from high mountains that then supported such abundant vegetation, whose soil is now so diminished in both quantity and quality as to 'grow just barely enough for bees' (*Criti.* 111c4–5). This hint at human practices of deforestation suggests the possibility that they exacerbated the geographic features primarily responsible for causing the erosion.[15]

Divine rule, human rule and ecology

So far I have argued that conceptions of divine rule can still make room for collective human politics in affecting the environment. To flesh out the forms of possible interplay between divine rule (theology), ecology and human rule (politics), I will consider a case which might seem least hospitable to such interplay: the plagues visited upon the Egyptians in the book of Exodus, followed by consideration of the relationship between a people and a land in both the Bible and Athenian and Platonic thought.

The traditional view of the plagues is that they simply represent the divine will and the divine plan – announced as it is at the opening of the episode. Having commanded Moses to repeat what God will command him, through Aaron, to Pharaoh, to let the Israelites depart, God immediately adds: 'But I will harden Pharaoh's heart, that I may multiply My signs and marvels in the land of Egypt' (Exodus 7:3).[16] Nevertheless, in a recent account, Rachel Havrelock suggests reading them in a very different way: as representing the causal interactions of humans and the environment. She writes:

> The divine acts recounted in this parasha are commonly called 'plagues,' yet the term is used only rarely in the exodus narrative (Exodus 9:14). In fact, the disruptions that devastate the land and attack the routine affairs of ordinary Egyptians are most commonly called 'signs.' The biblical terminology for the plagues challenges contemporary interpretations of unusual weather patterns, super storms, and declining species by suggesting that such 'plagues' of weather might be best understood as 'signs' that the current levels of human consumption have a deleterious effect on the health and stability of our environment. (Havrelock 2017: 99)

So, are the plagues direct divine intervention, or naturally determined signals of human impact on the environment? I want to suggest an intermediate view: that the succession of the plagues as 'signs' (*'otôt*), and the outcome of the whole episode, while clearly involving divine agency, is at the same time crucially responsive to Pharaoh's political decision-making. Pharaoh's decisions to take certain actions or inactions play a key role in determining whether or not the signs will be manifested and the environmental and human health woes experienced. We might note that the word *'ot* is also used for the sign of the covenant in the form of circumcision, for Shabbat (Sabbath) as an eternal sign and for tefillin (phylacteries) as a sign on the arm. In all of these other contexts, some kind of human partnership and action is required in order for the sign to function as intended by God. And this is exactly what we find also in the case of the plagues.

At almost every step of the plagues narrative, Moses is told to tell Aaron (his spokesman) to offer a choice to Pharaoh: it is up to Pharaoh to make a plague cease by letting the people of Israel go out of the land of Egypt (in fact, the request is that they be allowed to go out to celebrate a religious festival, and this description of their purpose is maintained throughout the story, even though it eventuates in their permanent departure). At almost every step, Pharaoh initially agrees to the demand (sometimes with conditions) and the plague ceases, but then Pharaoh reneges on his agreement and so the next plague is announced – sometimes with warning to Pharaoh that should he choose, he can still avert it, at other times not. It is his reneging on the agreement that triggers almost every plague.

At almost every step along the way, it is Pharaoh who – either hardening his heart himself (as described about half the time) or subject to God's hardening of it within him (the other half) – has the power to block the people from leaving. Now these contrasting descriptions, coupled with God's original announcement of his plan of hardening Pharaoh's heart, make it notoriously difficult to attribute responsibility to Pharaoh. Without attempting to resolve that major theological and textual issue, I would observe that while there is a problem about attributing responsibility between God and Pharaoh for the hardening of Pharaoh's heart, there is no such ambiguity about the responsibility for his eventual decision to let them to go. That is, it is not divine force majeure (as will happen with the parting of the Red Sea) but rather Pharaoh's decision and command that ultimately enables the Israelites to leave Egypt (Exod. 12.31). Indeed, the text continues that 'The Egyptians urged the people on, impatient to have them leave the country, for they said, "We shall be dead"' (Exod. 12.33), an interesting highlighting of the power of the people even under the monarchical rule of Pharaoh.

Thus, on my account, the plagues are not purely natural 'signs' of collective human biological agency, as it were. Nor are they purely divinely inflicted sufferings. Instead, the divine infliction is mediated by human political acts or refusals to act, and the plagues are 'signs' of a political ruler's stubbornness as much as they are of a simply causal relationship between humans and nature, or of a simply theological relationship between humans and God.[17] As Diana Lipton reflects on how the plagues might bear messages for today: 'We need to see the Nile as if it were the Jordan … we need to see ourselves as the Egyptians, suffering the economic and ecological consequences of our devastated natural environment; and we need to see the pollution of a river as something to be resisted and reversed at all costs' (Lipton 2017: 103).

Divine rule, human agency and the condition of the land

What is true of the plagues is true more generally of many Hebrew, and also Greek, accounts of the relationship between a people and their land. What may seem dramatically contrasting accounts – on the one hand, the biblical promising of a particular land to the Hebrew patriarchs and their descendants; on the other, the Athenian myth of autochthony (namely the mythical idea of a people being born from the very land to which they belong) and its appropriations by Plato – in fact make strikingly similar points at a more abstract level: that human action shapes the fertility or barrenness of the land, even when inscribed within broader theological accounts of divine agency as well.

The land in its original biblical promise is presented as plentiful and abundant – 'a land flowing with milk and honey' (Deut. 6.3), one which is more extensively described as 'a land of wheat and barley, of vines, figs, and pomegranates, a land of olive trees and honey; a land where you may eat food without stint, where you will lack nothing; a land whose rocks are iron and from whose hills you can mine copper' (Deut. 8.7–10). Note how these descriptions combine elements of food without toil (olive trees, perhaps figs and pomegranates), minimal effort (perhaps milk, honey, vines), and with agricultural labour and artisanry (wheat, barley; iron and copper), into a general promise of abundance. While this is not Eden, it is a land in primordially flourishing and presumably sustainable condition. However, its prosperity is conditional, as we will now see by asking two questions as before: first, what kind of relationship is this – is it a 'sign' of divine favor, or is there some more directly causal link between humans and land here? And second, where if any is the room for politics?

God promises that the land will be fertile for those who act ethically and keep the commandments – in Leviticus 26, God promises rain if the people follow the laws and commandments, rain that will make it possible for the people to dwell 'securely' (*lābeṭaḥ*) in their land (Lev. 26.3–5), and to enjoy peace that will be a direct grant from God. Security is respite from wild beasts and war, both of them portrayed as kinds of ravaging encroachment of the wild upon cultivated and civilized human society, threatening its children, domesticated animals and routes of passage and communication (Lev. 26.22). If the people keep the commandments, the land will flourish. But if they flout them, God will inflict consumption and fever, allow enemies to eat the seed that the Israelites have sown, and withhold the rains to prevent the trees and land from bearing food (Lev. 26.15–20), followed by further punishments if disobedience continues. Here divine agency works through disease, through allowing other humans to wreak their wills in war, and through the predictable effects of withheld rain.

Consider in particular the repeated adverb (*lābeṭaḥ*) or 'securely'. Whereas the image of an obedient people in a prosperous land was one of security – we might say stability or even sustainability – the image of the consequences of disobedience is insecurity, instability, leading to a lack of confidence in sustainability. This reminds us that sustainability (like security more generally) has a subjective dimension as well as an objective one. Insofar as it must incorporate not only psychosocial stability (as I have argued in Lane 2011/2012), but also, as we see here, justifiable confidence in the reliability or stability of a fertilely productive land (itself dependent on the environment), sustainability is vulnerable to any decisions – divine or human – that undermine that confidence.

We find the point directly applied to politics in Leviticus 25's injunction to observe a sabbatical year for the land, leaving it fallow every seventh year and eating only of its spontaneous produce (25.1–7), and a jubilee year in every fiftieth year (after seven cycles of seven years), in which that sabbatical practice is combined with a return to original property holdings of those who had lost them through debt foreclosure (Lev. 25.13–17). God here reiterates the promise that if the people keep the commandments, they may 'live upon the land in security (*lābeṭaḥ*)' (Lev. 25.18; see also 25.19). Here, the context includes especially the people's obedience to the laws of the jubilee year which enjoin a cancellation of debt evictions and a concomitant regulation of property leases in every year leading up to the jubilee – political measures if ever there were some. It is especially, and pointedly, such political failure or malfeasance that would undermine the ability of the people to enjoy the produce of the land 'in security'. If politics does not allow the land to enjoy its sabbatical and jubilee rest, God will ensure that it rests – by laying waste to it, destroying the civilization that is enforcing its excessive 'toil', or perhaps by allowing enemies and wild beasts to do the laying waste without hindrance.

A similar set of thoughts about the human drivers of ecological destruction are expressed by Greek authors. When humans act unethically, Hesiod explains in *Works and Days*, they are typically driven by *hubris*, an arrogant disregard of boundaries. Those same motivations are likely to express themselves in actions that are damaging to the environment, such that the environmental future of the land depends in part on whether human action has been viciously or virtually motivated. Similarly, Hesiod describes that those who observe justice (a political concept) enjoy peace rather than war, and so 'feast … on the crops they have tended' (231). While he goes on to describe the general plentiful produce of their land in this happy condition, it is clear that human political justice leads to peace which leads to the safe agricultural conditions that generate a flourishing condition of the land. In Aeschylus' *Eumenides*, the Furies – divinities whose commitment to perpetual cycles of vengeance for murder of kin, is in the course of the play subordinated to new procedures of justice instituted in Athens by the Olympian goddess Athena – initially foreshadow that they will damage the land with blight in their anger at being dethroned (778–94). But they are eventually persuaded by Athena to give up their anger and chant a hymn of blessings of fertility over the land instead (938–55), a set of blessings that Athena makes conditional on the citizens continuing to honour the Furies (now renamed Eumenides, or kindly ones) with their due (989–95) (West 1998).

What about the relationship between a people and a land in Platonic and other Greek thought? In Plato's *Critias*, the gods are each allocated their own lands to settle and oversee, with an initially extreme version of objectifying pastoral rule – they 'began to raise us as their own chattel and livestock' (109b7). Hephaestus and Athena jointly receive Athens, and 'they fashioned in it good men sprung from the land itself and gave them a conception of how to govern their society' (*Criti.* 109d1–2, or, more literally, gave them an idea of the constitutional order). Once again, divine tendance – even the most seemingly extreme form of pastoral rule –still involves a role for humans themselves operating according to a conception of politics.[18]

Such invocation of autochthony is most closely tied in Greek thought to its role in the origin myths of the Athenians. Its political significance there has been variously interpreted; when Plato proposes in the *Republic* that a 'noble lie' (*Resp.* 414b8–c1) be told to the proposed founders and citizens of the ideal city about their common origin as born of the same earth, this has been understood as aiming to cultivate fraternity and solidarity, while also reinforcing the political exclusion of metics (non-citizens resident in Athens) from the imagined *polis*.[19] But it is also of course crucial that autochthony creates a relationship between a particular people and a particular piece of *land* (or one might say, creates a people *through* their relationship to a particular piece of land). According to Plato, the land of ancient Athens 'alone in that olden time bore food fit for humans, wheat and barley, which are the finest and best nourishment for the human race, because she really was the mother of this creature' (*Menex.* 237e7–238a3). It does so inasmuch as the land gives birth to the humans who belong to it, with the land portrayed – in Plato's complex rendition of the Athenian outlook on politics in the *Menexenus* – as one's mother as opposed to stepmother (237e1–238a5).[20] It is interesting to compare the Roman author Pliny: despite an anthropocentric understanding of the figure of Nature, who, Pliny claims, appears to have created all things for man's purpose, it is still 'hardly possible to judge whether she has been more a kind parent to man or more a harsh stepmother' (*Natural History* 7.1, as translated in Rackham 1942).

For Plato, it is even the land itself – 'the earth hereabouts' (*Menex.* 237e2) – that is ascribed agency in establishing divine care for the Athenians: the land somehow prompts the allocation of special gods to 'rule and teach' this people (Athenians), training them in arts for daily needs and for military defense (*Menex.* 238b1–6). But notice once again that while under this general divine 'rule', the Athenian ancestors 'lived under a polity that they had made for themselves' (*Menex.* 238b7–8). While the gods teach agricultural and military skills, here the ancient Athenians are said to have been able to form their own constitution, one which whether called 'an aristocracy' or by some other name, involves 'government by the best men along with popular consent' (*Menex.* 238d1–2). With an autochthonous connection to their land, and some divine assistance (and general 'ruling' oversight), the Athenians are able to govern themselves. Here the land itself is credited with the power not only to nourish adequately (if with some toil as well as some pleasurably spontaneous foodstuffs), but also to bring about good theological (divine) interventions, and finally, to foster productive political agency.

In our own time, theological conceptions of nature are much contested and far from widely accepted, even as their grand narratives of the place of humans and nature – whether domination or stewardship, creation or exodus – continue to play an important role in the cultural imagination. The idea that it is only in the Anthropocene that humans have become like gods belies the many and subtle explorations of the relationship between divine agency, human agency and the condition of the natural world, that the ancient world, both Hebrew and Greek, affords. Whereas in ancient times the different spheres of human influence were spatially demarcated in relationship to the divinity or divinities apportioning them different plots of land, it is now temporally demarcated in geological time for the globe as a whole.

Unearthing the submerged role of politics in these accounts reminds us human agency can have a powerful effect on the natural order, whatever one believes about the role of divine agency in history and nature as well. We have always been more powerful than we might have believed, politically as well as biologically even if not always geologically, and have acted at least under or alongside gods even if not always 'as gods'. That power should give us pause as we seek to convert it into more ecologically conscious, and conscientious, forms.

CHAPTER 2
THE ECOLOGY OF THE *SIBYLLINE ORACLES*
Helen Van Noorden

Introduction

In his book, *Traces on the Rhodian Shore,* Clarence Glacken surveys 'Nature and Culture in Western Thought from Ancient Times to the end of the Eighteenth Century' (Glacken 1967). Texts discussed in the large section on the ancient world range from Anaxagoras to the apostle Paul, and Glacken rightly emphasizes the diversity of ancient ideas. Still, he finds three questions dominating the history of Western thinking about humans and the earth, of which the third is: how has humanity's spell on earth changed it from its hypothetical pristine condition?

One ancient text intensely interested in this question, but which does not make it into Glacken's survey, is the surviving collection of *Sibylline Oracles*, an intriguing blend of archaic Greek epic, Hellenistic Jewish and early Christian traditions whose central preoccupation is precisely the interface between the divine will, humanity and the environment. These oracles in Greek hexameters write themselves into the literary culture shared by all those educated in the archaic poets Homer and Hesiod, while the utterances themselves are attributed to the pagan Sibyl, the most authoritative prophet in the Roman world (Potter 1994: 77–8). Sibylline 'Books' were consulted at times of crisis by a select college of priests in Rome. In reality, these oracles are Jewish transformations of Sibylline prophecy (with Christian supplements), and as pseudepigraphic compositions they seek to appropriate the prestige of the Sibylline Books. Whereas only fragments survive of those pagan Sibylline oracles (for which see Alexandre 1856: 118–47), the extant *Sibylline Oracles* present 4,243 verses divided unequally into 12 'books' (and various fragments), whose composition and redactions seem to date between the second century BCE and the fifth century CE. In a recurring theme of this text, the Sibyl predicts environmental upheavals for various nations, and holds up the example of the Jews, whose land remains fertile as long as they worship the true God.

This chapter explores the ecological ideas that bind together disparate 'Sibylline' presentations of prehistory, ethics and eschatology. It argues that foregrounding environmental concerns offers a key to reading the prophetic idiom of this 'Sibyl', in particular as a focus for understanding how she relates to both her Homeric and Biblical backgrounds and makes them interact.[1] The *Sibylline Oracles* begin from the idea that the earth was divinely created as a moral order (versions of this point are to be found in books 1.8–25 and 4.13–17) and from there consider what Glacken takes as his third study question: how the earth reacts to human behaviour. To give just one example:

in one memorable tirade against *hubris* ('wanton violence, arrogance, outrage'), the Sibylline speaker asserts, in a metaphor perhaps recalling the divine punishment of Noah's generation: 'you brought on a full flood of evils' (5.234 *holon ... kakon kateklyssas*), and 'through you the beautiful folds of the world were altered' (*kosmoio kalai ptyches ēllachthēsan* 5.235).[2] In this latter phrase, a Homeric expression for hill country has been extended to the cosmos, in a move typical of this text, which reflects a contemporaneous Jewish emphasis on guardianship of the earth granted to mankind (see Marlow 2009: 83 for biblical references). Such fusions of Classical and Biblical elements in the Sibylline texture are particularly interesting in sections touching on the causation of environmental events.

As a text whose presentation of history and landscape fuses diverse ancient mythologies, the *Sibylline Oracles* bear comparison with Virgil's bucolic re-envisioning of disparate source-material in the collection known today as the *Eclogues*, a series of poems which Timothy Saunders and others have discussed in terms of environmental literature (see Saunders 2008 with further bibliography). In particular, Virgil's *Eclogue* 4, known by early Christians as the Messianic *Eclogue* for its prediction of the birth of a boy who will bring back the Golden Age, in its opening lines evokes multiple systems of time-reckoning alongside a Sibyl from Cumae in Italy (later to be immortalized as the location for Aeneas' Sibyl-led descent to the underworld in Virgil, *Aeneid* 6). The poet presents human history as a succession of eras: 'Now the last age of Cumae's prophecy has come; the great succession of centuries is born afresh' (*Ecl.* 4.4–5).[3] Virgil's blending of the Stoic theory of recurring events, Etruscan ages, ancient receptions of Hesiodic human races and the divine succession myth problematizes, even as it sets forth, the contemporary assertion of a turning point in history (Perkell 2002: 15). Several scholars have suggested a link between Virgil and the *Sibylline Oracles*, especially in reference to their visions of paradise (see section 'Prophecy *kata gaian*' below); whether or not this is the case, we may observe various ways in which both Virgil and the Sibyl co-opt the earth into their moralizing, and each section of this chapter will make comparison with Virgil's poetry to clarify the equivalent point about the Sibylline presentation of a divinely founded ecology.

I will consider, first, the basic presentation of 'the land' in the *Sibylline Oracles* (section 'Prophecy *kata gaian*') and its use of ecological metaphors to describe both mythical and historical events (section 'Thinking through the earth'). I then discuss the overlap in this text between natural disasters and acts of God to warn, punish or reward humans for their behaviour (section 'Co-opting the earth into the argument'). Finally (section 'The *Sibylline Oracles* as ecocritical literature'), bearing in mind the thoughts of Richard Kerridge (2014) on literary forms that might provoke care for the environment, I note how thematic and formal aspects of the *Sibylline Oracles* are mutually reinforcing if classified as an ancient example of ecocritical literature.

Environmental thinking is one of the areas in which the unusual blend of pagan and Biblical traditions in the *Sibylline Oracles* may be clearly seen and arguments made regarding the potential interest of this text for both classicists and theologians, let alone modern environmental literary critics. This chapter argues that the environmental focus of the *Sibylline Oracles* brings out the common ground between archaic Greek and Biblical poetry, whether the Sibylline speaker recasts Biblical ideas in Homeric terms or, conversely, resituates Homeric phrases and scenes in apocalyptic contexts. The combination of sources, both in the text's various eschatological scenarios and in its views of history, recasts ecological tropes as part of the Sibylline idiom.

Prophecy *kata gaian*[4]

At the most basic level, the *Sibylline Oracles* present themselves as focused on the earthly environment, reviewing its trials and tribulations from the remote past to the last days ahead. The Sibyl of book 3, agreed to be the oldest section of the extant collection (perhaps second century BCE, certainly before 40 BCE; see Collins 1997: 187–8), identifies herself as a relative of Noah; this gives her the authority to retell primordial history in a narrative that blends Genesis and Hesiod in its account of the first races after the flood. The Sibyl then explicitly rivals Homer's account of Troy as she 'predicts' its destruction. Finally, she addresses a range of communities, from Greeks to Romans to Jews, and cities in Asia Minor, Egypt, Libya and Gaul, in a series of oracles of 'weal and woe'. Other books of the *Sibylline Oracles* seem to expand on book 3 in their scope and schematization of history, or to complicate it with a view of history as heading towards a second global catastrophe in the form of fire (see Lightfoot 2007: 128–9), but the treatment of the earth for the most part remains within the same parameters.

It is worth highlighting at the outset the poetic/rhetorical character of Sibylline descriptions of the environment. First and foremost, landscape is presented in the binary terms of biblical imagery, which offers 'ideal' and 'unideal' archetypes of experience (Ryken et al. 1998). For example, a fertile valley is opposed to a dark forest or wilderness, and in the realm of plants, green grass or a vineyard are desirable features, while thorn and weeds are to be avoided.[5] A stronghold with well-built foundations is opposed to a wicked city; the tranquil spring is contrasted with an overflowing river; weather is either supremely calm or tempestuous. Also notable is the use of merisms, two-fold polar expressions to designate a whole, most familiar from biblical poetry. 'Heaven and earth' designates the cosmos (e.g. at 3.786); 'animals and men' its inhabitants (see further Lightfoot 2007: 445). Other such expressions are familiar from Homer; 'cities and rich fields' (*Odyssey* 8.560) is found at *Sib. Or.* 3.582 and 3.750 to designate the whole earth (Buitenwerf 2003: 200).

Part and parcel of the Sibyl's land-focused perspective is that ultimate blessing, curse and signs of the day of judgement are all vividly portrayed in terms of miraculous transformations of the earth. Most famously, *Sib. Or.* 3.777–9 invokes the eschatological kingdom as one in which 'all paths of the plain and rugged cliffs, lofty mountains and wild waves of the sea will be easy to climb or sail'. The calmed sea in this vision is a Graeco-Roman concern, not found in the Biblical parallel, Isa. 40.3–4 (Buitenwerf 2003: 290). Another, less consoling vision of eschatological upheavals envisages that 'one day the sea will be dry ... great Asia then will be water ... and Crete a plain' (*Sib. Or.* 5.447–9).

We have already seen, in the Sibyl's outburst against mortal *hubris* altering the 'beautiful folds of the earth' a negative expression of this notion about transformation of the earth at the eschaton.

In the *Sibylline Oracles*, the basic view of a blissful existence is of the earth overflowing with wine, honey, milk and corn (e.g. 3.619–22), a perspective that reflects a Jewish tradition going back to Exodus (e.g. Exod. 3.17). However, a similarly bucolic vision of paradise is found in Virgil's *Eclogue* 4, which builds up to a vision of nature in technicolour following the birth of the child who will bring back the golden age. It is asserted that every land will bear everything, so that there is no need for trade or agriculture (*Ecl.* 4.38–41); a picture amplified by a depiction of tamed or absent wild animals (22–5) and spontaneously self-dyeing sheep in the fields (42–5). The pastoral terms of the vision of the 'last age', which dominates the poem, do help to situate this unusual poem within the *Eclogues*, even though it has relatively little pastoral content compared to the other poems in that collection. Virgil's introductory exploitation of the voice of a Cumaean sibyl long ago prompted Roland Austin to note that the rare volume of spondees, 7-line patterns and end-rhymes in this *Eclogue* mimic the tone and style of Sibylline prophecies (Austin 1927). In particular, Eclogue 4's pastoral vision shares several features of content with a later section of the Isaiah-derived Sibylline passage, mentioned above, about the eschatological kingdom on earth (also notable for its end-rhymes in the Greek):

> Wolves and lambs will eat grass together in the mountains.
> Leopards will feed together with kids.
> Roving bears will spend the night with calves.
> The flesh-eating lion will eat husks at the manger
> like an ox, and mere infant children will lead them
> with ropes. For he [God] will make the beasts on earth harmless.
>
> (*Sib. Or.* 3.788–93, drawing on Isaiah 11)

The possibility of Virgil's having knowledge of a tradition of Jewish Sibylline Oracles has been much discussed (for a sceptical review of the scholarship see Collins 1997: 192–7) and a two-way connection has recently been mooted (Neutel 2015: 63–4).

One less often noted detail may be used to highlight the perspective on paradise held in common by *Eclogue* 4 and the *Sibylline Oracles* – the notion of the land as a miraculous 'gift'. While the early poems of Virgil's bucolic collection play with various ideas of gifts of and from parcels of land (*Eclogues* 1 and 2), the concept is presented at another level both in *Sib. Or.* 5.331, where fertile Judaea is described as God's 'special favour' (*procharisma*) for all mortals, and also in Virgil's idea of uncultivated nature offering its 'little gifts' (*munuscula*) to the little child, agent of the golden age (*Ecl.* 4.18). The impact of this idea in Virgil's poem emerges by contrast a few lines later, as it is strikingly emphasized that the golden age will not be fully realized immediately, and the *status quo* is one of mortals virtually attacking the environment to lead their lives within it:

Traces, though few, will linger yet of the old deceit,
commanding men to tempt (*temptare*) Thetis [the sea] with ships, to encircle
towns with walls, to inflict (*infindere*) deep furrows on the Earth.

(*Ecl.* 4.31–3)

Here, the verb *temptare*, which has overtones of 'assault', assimilates sea trade to war; the juxtaposition of city defences and agriculture makes the furrows cleaved in the earth comparable to the dividing walls. This detail in its turn takes us back to the *Sibylline Oracles*, where the Sibyl envisages the earth restored for the righteous, undivided by walls or fences, and lives 'in common' (e.g. *Sib. Or.* 2.319–21, 8.208). Such an emphasis on lack of private property is a feature found only in Roman visions of the golden age (after Virgil, see e.g. Ovid, *Metamorphoses* 1.89–150). We may detect its influence also in a later (apparently post-Neronian) book of the *Sibylline Oracles* that reflects in a more sharply sarcastic vein on the Roman love of ownership:

If the huge earth did not have its throne
far from starry heaven, men would not have equal light
but it would be marketed for gold and would belong to the rich,
and God would have prepared another world for beggars.

(*Sib. Or.* 8.33–6)

The basic focus on the earth in later sections of the *Sibylline Oracles*, then, is arguably at least as Roman as Greek and Biblical, and comparison with Virgil *Eclogue* 4 points up common ground between the two texts in their views both of paradise and of contemporary or everyday treatment of the earth.

Thinking through the earth

A rather more oblique strand of Sibylline thinking 'in an ecological vein' is worth outlining at this point, with Virgil's bucolic visions in view, and this is the striking use of environmental metaphors, which deserve more attention as a feature of the Sibylline idiom than they have hitherto enjoyed.[6] Virgil's *Eclogue* 6 is often paired with *Eclogue* 4 as a 'grander' poem than the others in the collection; it presents an account of Silenus, captured by shepherds, singing to them a history of the universe. The poem begins by exploiting a Classical tradition of bucolic metaphors for poetic choices, expressed as a choice of musical instrument – the poet will sing 'on a slender reed' (*Ecl.* 6.4–5) – but the content of Silenus' song, which shades into the voice of the bucolic poet, is nothing less than an account of the world's creation (6.31–40), followed by a series of well-known mythological tales, here presented in terms of the protagonists' interactions with the environment, fantastical and otherwise (41 to end). This focus on elements of transformation of the natural environment is an angle on classical mythology later developed in great detail by Ovid's epic, the *Metamorphoses*.

On the level of metaphor, such a viewing of human history 'through the earth' is a mode of speech familiar to readers of biblical books such as Amos, Hosea and Isaiah (see especially Marlow 2009). While some scholars have worried about the use of nature imagery in the Bible as evidence of a (further) anthropocentric bias, and as an 'instrumental' use of the world for thinking about human reality, the basic use of such metaphors in fact stems from the human sense of connection to the natural world (as Marlow 2009: 124–5 emphasizes). Although the Sibylline speaker does make use of animal metaphors, so common in the oracular discourse of many ancient traditions, in prophecies about individuals, it is her plant and weather metaphors which produce a strikingly long-distance lens on large-scale man-made events. In the *Sibylline Oracles*, in my view, the impression given is that the Sibyl has such a distant perspective on both past and future that she as a viewer cannot divide organic from non-organic disturbances of the earth.

The first example I will focus on is the mythological event of the sack of Troy. It is introduced (in the Sibylline style of *post-eventum* prophecy) with the 'evidence' [*tekmar*] for Phrygians of the destruction of

> the abominable race of Rhea, an everflowing surge in the earth (*en chthoni kuma /aenaon* [translation mine]), flourishing with unthirsting roots (*rhizēsin adipsētoisi tethēlos*), [which] will disappear root and branch (*autopremnon*) in a single night.
>
> (*Sib. Or.* 3.402–4)

In the context of Book 3, this 'race of Rhea' comprises the inhabitants of a city described in terms that fit the city of Troy, but which is then formally identified as Dorylaeum in Phrygia. The story of Rhea's offspring has already been related earlier in the book, including the fact that she sent her baby Zeus secretly to Phrygia to avoid the wrath of his Titan father Cronus (3.140). Phrygians are therefore at 3.401–2 being viewed typologically as connected to Zeus, and (in the Jewish appropriation of the Sibylline voice) are negatively presented apparently as those who worship the pagan gods. They are glossed as *kuma*, a term which refers to anything swollen, used in Classical poetry with the sense of both a billowing wave and an embryo. The epithet 'everflowing' and the paradoxical emphasis 'in the earth' point to the former meaning 'wave', here, while the adjective *tethēlos* (from *thallō*, 'bloom' or 'sprout') in line 403 points to the latter, organic, meaning, the plant metaphor then developed with the oxymoronic phrase 'unthirsting roots'. The use of an environmental lens on such an event in the *Sibylline Oracles* is expanded in a later section of the collection, Book 1, where these phrases are made over into a more general symbol of the divine destruction of cities and contextualized with a description of God's worldwide flood, which is described as 'strange water' (1.183: *epēlyton hydōr*). As Lightfoot 2007: 395 *ad loc.* notes, the adjective 'strange' implies an alien imposition on the earth, in this context working to rival the monstrosity of the evil 'race of Rhea'.

The destruction of Ilium/Troy itself is prophesied in lines following soon in Book 3 (414–17), which refer to the birth of Helen, the cause of the Trojan War, as a very beautiful 'shoot' (*ernos*) which will 'sprout' (*blastēsei*) in Sparta and bring lamentation in its wake. Such a metaphor is a common one for human offspring (cf. 'family tree') but here inverts the positive sense of beloved and flourishing youth with which it is associated in pagan Greek epic (*LfgrE s.v.* ἔρνος). Buitenwerf (2003: 229) rightly compares the plant imagery of a Jewish addition to a pagan oracle a few lines earlier, at 3.396–7: 'leaving one root, which the destroyer will also cut off / from ten horns, he will sprout another shoot (*phyton ... phyteusei*) on the side.' Such an image, based on Dan. 7.7ff., the vision of the fourth Beast with ten horns and an additional horn that arose and destroyed three others before being overcome by the kingdom of God, is perhaps the inspiration that enables the Sibylline speaker to imply a similar overthrow of pagan races with the application of a similar plant image; for those who know Daniel 7, she does not need to spell out the ending.

Moving to the historical plane, in Book 4 we find a plant metaphor used also for the Macedonian wars (214–168 BCE): 'a great Italian war will bloom' (*anthēsei* 4.103), while alarming weather terms are employed to describe both Italy's war on Jerusalem (*thuella*, 'squall' 4.115) and, in Book 5, a prediction of the return of Nero (*cheimeriē pnoiē*, 'wintry blast' 5.375). Homeric similes have made familiar the comparison of both an army and an individual to features of a storm (e.g. *Iliad* 19.357–61 of the army; 11.297 and 12.40 of Hector), but the idea of war as 'blooming' is rather more striking; it plays against an earlier Sibylline vision of war with an emphasis on the barrenness of the earth in times of war. When in Book 3 the Sibyl shifts from predicting slavery for the Greeks (in language reminiscent not only of Deuteronomy 28 but also of the Trojan hero Hector's gloomy predictions for his war-torn wife Andromache in Homer, *Iliad* 6) to a description of forthcoming world-wide war and pestilence, she glosses these phenomena in terms of a transformed earth:

God will make a **great bronze heaven** on high
and cause drought over the whole earth, and make the **earth iron**.
(*Sib. Or.* 3.539–40)

The emboldened phrases are found in Leviticus (26.19) and Deuteronomy (28.23), but in the *Sibylline Oracles* the Iliadic build-up to these lines may prompt a recollection of Homer's reference to a bronze sky during a battle scene (*Iliad* 17.424–5 – 'bronze' perhaps because of a shower of arrows, or the stirred-up dust). The Sibyl has transformed the Homeric and biblical perspectives into her own ecological idiom of lament at how war destroys the environment, albeit asserting its divine origin. Elsewhere, she highlights the effects of war on the earth as a form of environmental pollution: 'The bloody ocean will be filled with flesh /and blood of the senseless, from evil war' (5.472–3).

Finally, and most strikingly of all, an ecological metaphor is deployed by the Sibylline speaker in eschatological prophecies, applying to the future images of sowing and reaping which were used in (e.g.) Hos.8.7 and 10.12–23 to predict consequences of the behaviour of Israel, faltering in its commitment to the covenant. For the Sibyl, this future is 'when the Hebrews the ill harvest reap' (*Sib. Or.* 1.387, by implication deservedly) and she likewise prophesies that when women are barren, 'the harvest of articulate men has come' (2.164: *ephy to theros meropōn anthrōpōn* [with Collin's translation]). As Lightfoot notes on this latter line, the use of the past tense (intransitive *ephy* – 'has come') here indicates completion; and while the more usual word for harvest in this context is *therismos* found in Mt. 13.30, 39 and elsewhere, *theros* in this usage has a classical lineage; the reaping metaphor in *Sibylline Oracles* 1 recalls the 'harvest of woe' from Aeschylus' *Agamemnon* 1655, and the more universal image in *Sibylline Oracles* 2 calls to my mind the 'woeful harvest of men' image of destruction running through Aeschylus' *Persians* (e.g. 822).

Such Sibylline developments of environmental metaphors showcase the way in which she blends biblical and pagan expressions (especially to describe times of suffering) and reorients them towards an eschatological future. The 'environmental lens' is a distinctive feature in the Sibyl's prophetic idiom, which markets itself as a broader and longer perspective than all others including Homer. Such ecological language helps to emphasize the Sibyl's more cosmic vision even as it also brings out the 'common wisdom' between the traditions on which she draws.

Co-opting the earth into the argument

Beyond environmental metaphors, we may see the Sibylline speaker 'co-opting the earth into the argument' in the Sibyl's perspective on *natural* events (especially disasters) as responses to man-made events and mortal behaviour. Most notably, the eruption of Vesuvius (79 CE) is presented as a punishment on the Romans for their destruction of Jerusalem (70 CE) (*Sib. Or.* 4.130–6). In the text this event is connected, at least implicitly, to the 'wrath of God', but the agency of such acts of 'retribution' deserves more comment and will be discussed further below in this section; for the moment, what should be noted is the overlap here between two common phenomena in these oracles; the law of the *talio* ('retaliation in kind', seen e.g. in the prophecies of Rome paying back many times all that she has caused Asia to suffer, *Sib. Or.* 3.350–5) and what the Victorian critic John Ruskin termed the 'pathetic fallacy' ('reading' in the status of the environment a sympathy with or reflection of the feelings of human agents).

Such overlap derives from the *Sibylline Oracles*' presentation of the earth as a 'created moral order' (as Fretheim 2005: 163–5 describes the world as presented in the Old Testament), whereby the prosperity of the land is tied to the ethical behaviour of its inhabitants. This notion is seen whenever the Sibyl differentiates the 'pious' from the 'wicked' in her predictions of the eschatological kingdom on earth. In describing transformations of the earth in the end-time, described above, the Sibyl concludes: 'all peace will come upon the land of the good' (3.780). Elsewhere, she threatens that until mortals cease to worship idols, 'all will remain unsown and unploughed' (*asparta ... kai anērota* 5.276). The use of these adjectives in this context overturns a key association in Classical portrayals between the 'unploughed' prosperity of the earth and the golden age, reflected e.g. in the prophecy of the returning golden age in Virgil, *Ecl.* 4.18–20: **nullo** *munuscula **cultu**/ ... tellus/ ... fundet*, 'the earth **untilled** will pour forth its little gifts' (my emphasis in bold). For the Sibyl, only 'the holy land of the pious' will bear milk and honey (5.281–3).

So far, this feature of the *Sibylline Oracles* may be thought closer to Biblical precedents than Classical pagan ones. An assertion of continuity between social justice and the wellbeing of the land (and their opposites) is found in various forms in the Hebrew Bible. This is expressed as a condition of obedience e.g. in Lev. 26.3–5 where if the Israelites follow the Lord's commandments, their land will be fertile and peaceful. It is also found as an indictment of injustice in Isa. 5.8–10, Amos 5.11 and 8.4–10 where those selfishly acquiring land are threatened with its barrenness (see Lane in this volume pp. 13–23 for further discussion). However, the idea is common to several ancient traditions. For example, the notion lies at the heart of the *Oracle of the Potter*, a Hellenistic Egyptian prophetic text purporting to be a revelation from the potter Khnum, who fashioned the world, conveyed to and written down by Amenhotep (a pharaoh of the 18th dynasty). The prophecy asserts that Egypt will be barren and dry until its inhabitants become pious. Likewise, classical Hellenistic poets could have derived this idea directly from archaic Greek epic (Erler 1987). The notion that justice is rewarded in the experience of the land is found already in both Homeric and Hesiodic poetry in the context of presentations of ideal kingship. In Homer, *Odyssey* 19. 109–14, Penelope is compared to a just king under whom the land flourishes (e.g. the fields bear barley, the trees are heavy with fruit). Hesiod's *Works and Days* (225–36) suggests that those who rule justly could win a life like that of the golden race; for their community, the earth bears 'the means of life' in abundance, whereas the city of the wicked will suffer 'famine together with pestilence' (v. 243). The latter image in Hesiod however is part of the 'justice' dealt out by 'the son of Cronus [i.e. Zeus]' (v. 239), so we come back to the idea, to be amplified in this section, that a moral order is maintained through divine interventions in the earthly environment.

When asking if there are any distinctive emphases in the Sibylline presentation of a 'created moral order', it is worth comparing the way in which Virgil amplifies the traditions he has inherited. *Eclogue* 4 offers an extreme positive form of 'pathetic fallacy' to describe the earthly impact of man-made events, as it describes in stages the joyful response of nature to the birth of the boy who will bring back the golden age. Immediately, snakes and poisonous plants will perish (4.24–5); as the boy learns to read and understand virtue's potential, fields grow yellow with wheat, grapes will hang from thorn bushes and tough oaks will sweat dewy honey (4.28–30). The advent of a fantastical level of peace and abundance, a dramatic extension of seasonal transformation of the earth, is tied to the arrival and maturity of the long-awaited child.

In such a move, which identifies an entire era of earthly history (another 'golden age') as dependent on the birth and actions of an individual, Virgil seems to be a pioneer. To be sure, the poem attributes to Sibylline prophetic tradition the idea of the 'last age' (4.4: *ultima Cumaei venit iam carminis aetas*, literally 'now is come the last age of Cumaean song'), before asserting that 'the great line of centuries begins anew' (4.5). One could therefore gloss the Sibylline prophecy either as having predicted the worst age in the series (as Servius' commentary seems to understand it), or a reborn golden age, or the kingdom of God. Still, an emphasis on the connection of the advent of paradise on earth with an individual human 'agent' is not found in the *Sibylline Oracles* that survive. While there are several 'saviour figures' in the extant Sibylline collection, the Sibyl appears to show more interest in the negative end of the spectrum. These oracles repeatedly, for example, foretell the return of Nero, who will bring war and thunderstorms in his wake (*Sib. Or.* 5.370–80), before peace is restored for the wise – an interim future stage that Virgil's poem mentions (*Ecl.* 4.85: *erunt etiam altera bella*), but does not dwell on.

Equally striking in the *Sibylline Oracles* is the expression of a moralized pathetic fallacy in descriptions of the earth and sea reacting to mortal arrogance. These reactions take several forms. One is evasion. The Sibyl fears that the day may come

> when the sun is setting so that it never rise again,
> remaining to be plunged in the waters of the ocean,
> for it saw the impious wickedness of many men.
>
> (*Sib. Or.* 5.477–9)

The same idea is found in the first extant section of the *Oracle of the Potter*. However, another form of moralized environmental reaction described in the *Sibylline Oracles* is precisely the opposite of avoidance; the sea is imagined as an agent of punishment when it is envisaged as 'spontaneously encroaching' (*Sib. Or.* 5.127: *ap' automatou epibas*) to flood the 'grievous' land of Lycia (crime unspecified). The most intriguing example of such a sense of agency given to the environment presents the sea about to flood the Titans who rebelled against heaven, in a semi-disguised version of the assault on heaven implied by the building of the Tower of Babel:

> The ocean's mighty surge shall come on them,
> The waters' rage. The Lord of Hosts in wrath
> Will stem its flood, because he undertook
> Never again to deluge wicked men.
>
> (*Sib. Or.* 1.315–18)

The sea is about to act to punish, but God restrains the surge, recalling his promise to the descendants of Noah never to repeat the Flood. While there are various connections in pagan literature between the Giants or Titans and a flood that may be recalled by this passage (documented by Lightfoot 2007: 417–18), what is striking here is the image of a destructive element recalled from its purpose by a greater power. There is Homeric precedent in the famous episode of the river Scamander thwarted by the Olympians from taking his revenge on Achilles (*Iliad* 21.233–382). However, Homeric gods are famously whimsical in their interventions; in the *Iliad*, at least, divine rewards and punishments for individuals or even cities are very rarely prompted by moral considerations as opposed to whether human actions have pleased or displeased a particular divinity. It would not be in the Sibyl's interest to emphasize this aspect of her pagan legacy,[7] and we must count this incident a one-off. Indeed, Lightfoot finds the 'freak' potential flood incident in *Sibylline Oracles* 1 a symptom of 'slight bungling' in its mix of Genesis with later Jewish exegesis and classical mythography (Lightfoot 2007: 417 with further references).

There are, however, other examples which, while perhaps purely dramatic expressions, raise the question of whether the environment in the *Sibylline Oracles* is purely an expression of God's will or to what extent the earth or sea or sky are envisaged as independent moral agents (see Marlow this volume pp. 119–136 for further discussion). How should we read, for example, the emphasis that *gaia kai autē*, 'earth itself also' will drink the blood of dying men (3.696), or the voice implicitly attributed to the land which consumes human remains, *autē* ... / *kēryssousa talaina mysos myriōn anthrōpōn*, 'itself ... / wretched proclaiming the defilement [my term] of innumerable men' (3.647–8)? Classicists may be reminded of the overpopulated earth complaining to Zeus about her burden, an event that, according to the epic poem *Cypria* (frg. 1), caused Zeus to start the Trojan War to reduce the human population. However, where the Classical example envisages the Olympian god as an audience of a complaint, the cases of 'independent action' granted to nature in the *Sibylline Oracles* seem to be reinforcing the flow of the prophetic narrative rather than countering it or provoking a new set of actions.

A more apt precedent seems to be the use of non-human creation 'mourning' in Amos (e.g. 1.2), Hosea (4.3) and Isaiah (24.1–13) to proclaim God's anger (see, e.g., Marlow 2009: 201, also 152 on Amos 8.8 as God's choice to speak through his cosmic agent when the people have not listened to the prophets). So too in the *Sibylline Oracles*, actions attributed to aspects of the environment tend to occur at the climax of a series of prophecies about how divine anger or judgement will make itself known. The most dramatic example comes at the conclusion of Book 5, the past-tense narrative of a divinely-instigated 'battle of the stars', in which finally 'Heaven itself was roused (*ōrto men Ouranos autos*) until it shook the fighters [the constellations]. / In anger (*thymōtheis*) it cast them headlong to earth' (528–9).

Such phrases as these, attributing independent agency to parts of nature seem, moreover, to be exceptions in this text. As a rule, reactions of the natural environment are presented in the *Sibylline Oracles* in closer connection with the divine will, if not directly as manifestations of it. Most commonly, God sends punishment or reward through nature. A good example comes from the section agreed by scholars to be the oldest of the collection, and as such, part of the foundational layer of a long tradition of Jewish Sibylline prophecy. The version of the Tower of Babel adopted by the *Sibylline Oracles* recounts how God, before imposing different languages to end this mortal building project, sends winds against the Tower to topple it:

> But when the threats of the great God are fulfilled
> with which he once threatened men when they built the tower
> in the land of Assyria ... they were all of one language
> and they wanted to go up to starry heaven.
> But immediately the immortal one imposed a great compulsion

on the winds. Then the winds cast down the great tower
from on high, and stirred up strife for mortals among themselves.

(*Sib. Or.* 3.97–103)

It has been noted that divinely-sent winds are likewise the cause of the tower's downfall in Jub. 10:26. In the Sibylline description of the project's failure, however, oracular style ('when … then') frames the passage [although not straightforwardly], and the language of nature operates first literally and then metaphorically (the winds 'stirred up' strife – *erin ōrsan*). There is a similar use of a weather term as metaphor when the Sibyl describes (5.507–8) divine retribution for the Ethiopians who destroy 'the great temple of the land of Egypt': God will 'rain down' (*brexei*, 'inundate') a terrible wrath (form unspecified). We may want to think of such phrases as extensions of the image of God as controller of tempests, a notion found in the *Sibylline Oracles* in narrative (e.g. *Sib. Or.* 1.32–5) as well as epithets (Lightfoot 2007: 546–7).

Elsewhere, environmental upheavals are presented as forms of divine retribution, often in conjunction with war (i.e. 'man-made' destruction). One clear example, described in the above section 'Co-opting the earth into the argument', is the framing of the eruption of Vesuvius as punishment on the Romans for destroying the Jerusalem temple; another such punishment for destroying the 'house of the Immortal' is predicted to fall upon the 'daughters of the West' (probably the Romans and perhaps another nation) in the form of war, famine and pestilence (*Sib. Or.* 3.328–33).[8] Cyprus and Sardinia 'amid great storms of winter and blows (*plēgais*) of the holy God throughout the depths of the sea will sink in the wave' (3.477–9). Such a collocation of threats, effectively redescribing the storms as actions of God, is typical of the *Sibylline Oracles*; elsewhere God's wrath is glossed as a pestilence (*loimos* 3.633). On Babylon will fall a 'tumult' (*alalagmos*, literally 'shouting') and an 'affliction' of God (*plēgē*) (3.304–5) 'because it destroyed God's great Temple' (3.302). Already in Biblical prophecy (e.g. Isaiah 13) the downfall of Babylon is presented as a divine action. However, the abstract language of this last example from the *Sibylline Oracles* is typical of the oracular manner of the work – it is not clear from the context whether a literal 'plague' is meant or whether 'affliction, blow' is to be understood more figuratively; it has been noted that the term is elsewhere used in this text both of natural disasters (as for Cyprus, cited above) and, it appears, of destruction by enemies (3.501, 505 with Buitenwerf 2003: 215 *ad* 3.303–6).[9] Some sections of the *Sibylline Oracles*, however, present a series of cities that will suffer the same form of destruction, such as flood or earthquake, and in these cases, one might think that memories of literal and historical environmental destructions are being made over into prophecy.

In such prophecies of natural destruction, the Sibyl seems to sit between pagan and Jewish apocalyptic traditions and arguably acts as 'interpreter' of their common insights. When the Sibyl comes consciously close to Homer in her prophecy (3.401–9) of the fall of 'the city of the earthquaking landshaker' (phrasing which recalls Troy, whose walls were built by Poseidon, although the Sibyl identifies the city with Dorylaeon), she attributes its destruction to a 'time of earthquake' (3.408 *kairos … enosichthōn*) rather than an earthquaking god.[10] This might be seen as a move to de-mythologize the pagan epic tradition of wilful divinities (cf. my observation above on *Sib. Or.* 1.315–8). Still, the power and intrinsic connection of the divine in nature is clear; the Sibylline view of the world is one that sees the destruction of an ancient city by an earthquake as 'evidence' (*tekmar* 3. 401) of the divine will in the world, whether that is understood as punishment in itself or a portent of something to come.[11]

The resonances of earthquake as portent, in particular, appear to have shifted both within pagan antiquity and within the Jewish tradition. Susan Satterfield has recently argued that while prodigies such as earthquakes and comets were widely seen in antiquity as signs of divine anger, it was possible for Augustus in 17 BCE to transform them into positive 'signifier[s] of power and pre-eminence' (Satterfield 2016: 344). Lightfoot, meanwhile, notes the transformation of earthquake from an Old Testament 'theophany motif' into an eschatological sign in apocalyptic writing (Lightfoot 2007: 445 with further references). Lightfoot distinguishes this tradition, which signals that 'the times are out of joint', from the Classical use of portents as prompts to remedial action, but I would argue that in the *Sibylline Oracles*' treatment of earthquakes, the two traditions are brought very close together.

For those seeking to understand the theology behind such Sibylline prophecies, it may be too much to expect to find a single, coherent, underlying idea of the relationship between God and the natural world in the *Sibylline Oracles*, given the range in date and chronology of the oracles' composition and redaction. In terms of God's place in the world, Iliadic formulary is taken over for the repeated description of God as 'living in the aether' (e.g. 5.298), while lines such as 5.344–5 on the crash of thunder as the voice of God above may be felt to echo Psalm 29 about the power of the voice of God, felt throughout nature. One intriguing statement in the voice of God associates (although does not equate) the divine body with parts of the cosmos, as in Isa. 66.1 (and Mt. 5.35): 'the earth my footstool, I am girt with air' (*Sib. Or.* 1.139). Parallels for this conception in other ancient cultures, especially Greece and Egypt, and in particular the Hellenistic poem *Testament of Orpheus*, suggest that the Sibyl is here reformulating pagan material in a tradition of Judaism which 'tolerates a moderate degree of anthropomorphism' (Lightfoot 2007: 384–6). This trend appears to be combined with the self-consciousness of this text as *written* eschatological prophecy in the prediction (3.81–2; cf. 8.233 and 8.413) that God, dwelling in the aether, will roll up the heavens like a scroll (*biblion* 3.82), an expression whose vocabulary combines Isaiah 34.4 and Revelation 6.14 (see Lightfoot 2007: 235–6 n.98).

This section has explored how the *Sibylline Oracles*' treatment of the causation of natural events (as responses to human behaviour) fuses Biblical and Classical pagan perspectives on the work of the divine in creation. While the result of the blend of traditions is perhaps not as striking as that in Virgil's *Eclogue* 4, the *Sibylline Oracles* intriguingly press the boundaries between literal and metaphorical weather terms in extending archaic formulations (of natural disasters or aspects of the divine) in the direction of Biblical imagery. In the final section below, I will suggest a way of classifying the *Sibylline Oracles* that not only factors in the use of Homer and the Bible as shared literary cultures but also offers a means of uniting the form and contents of this text.

The *Sibylline Oracles* as ecocritical literature

As I have noted above, the prevalence and deployment of prophecies of environmental upheaval in this Sibylline collection seem to fall between the emphasis of Judaeo-Christian apocalyptic on such portents as signs that 'the times are out of joint and headed for a change or purge' (so Lightfoot 2007: 445) and the Classical and Biblical prophetic emphasis on such events as warnings of what is to come should mortals not follow the right path. This literature may therefore be counted both 'ecological' and 'critical' in the sense of wishing to stage an intervention in the history of mankind. In this final section I wish to put a new accent on the Sibylline blend of theology and ecology by noting that their peculiar form of stop-start prophecy and recycled elements of multiple ancient traditions answers surprisingly well the modern call for 'ecocritical' literature through which the urgency of the environmental crisis can be articulated.

The drive behind the disruptive form of modern 'ecocritical' literature is the emphasis on humans as part of larger, worldwide and chronological processes, according to which we should disrupt the security of a familiar anthropocentric narrative point of view. Such critique, associated with the 'new materialism' movement, is felt in literary uses of a 'collaging' effect: 'dispersed, ventilated, fragmented, multivocal, dialogical forms without stable narrative viewpoint', or narrative involving 'breaks, interruptions and switches of viewpoint and register offer[ing] possibilities for the representation of ecological relationships that go beyond the range of local place and individual perception' (Kerridge 2014: 368–9). The *Sibylline Oracles* are nothing if not a 'collage' that brings together numbers of different oracles from different periods and places, resulting in abrupt switches of time-frame and focus. Their outlook has been termed 'internationalist' but the drastic means by which they offer a perspective that goes beyond the local seems designed also to upset any sense of a stable standpoint.

Kerridge (2014: 368–9) cites Harriet Tarlo for the idea that 'found poetry – consisting of text quoted from a variety of sources – is a genre especially suited to environmental concerns.' The 'recycling' of found text introduces questions about ownership and public space and 'foregrounds the question of whether literary culture can be regarded as a public commons, like the atmosphere.' Applied to the *Sibylline Oracles*, this works remarkably well; in their metrical form (hexameter) and their structuring content (world history presented as a moral decline in human generations from the first to the last) they ape Homer and Hesiod, the shared literary culture behind this text. The proviso is that the persona of the Sibyl actively lays claim to this culture, when she claims that a myth famously told by Homer, the fall of Troy, was taken from 'my words and meters … my books' (3.424–5). In reality, of course, it is the other way around, but while the Sibylline poet might on one level be called a 'magpie', borrowing and combining material and verse-elements from a range of pagan and Jewish sources, the development of the Sibylline tradition equally expresses the common and continuing importance of these stories and literary forms for communities of Hellenistic Jews/Jewish Hellenes. In this context, the urgency of the Sibylline prophetic voice reinforces the idea that both Greek and Biblical culture and myths deserve the attention of all.

Along with the point about the 'recycled' nature of the oracles in the Sibylline collection, one might argue that that the fragmented, generic nature of this text resonates well with our own contemporary emphasis on absence of coherent narrative about a world in crisis. A mid-first-century BCE parallel for this feature is again found in Virgil's *Eclogues*, this time with the emphasis on their (post-Virgilian) title, which means 'selections'. The collection is one of variety and of 'chips off the writer's block', and the poems variously allude to Octavian's re-allocations of farm land in the aftermath of the civil war. In particular, *Eclogue* 9, which presents two farmers walking to town singing half-remembered snatches of the songs of Menalcas (taken in antiquity to be a mask for Virgil), operates with a 'poetics of fragmentation' (this and the preceding description of the collection are quoted from Martindale 1997: 120). Martindale sees this Virgilian collection as a meditation on the necessary interweaving of aesthetics and politics and the ability of poetry to reflect and address worldly concerns.

We have seen that the basic presentation of the interaction between humans and the land in the *Sibylline Oracles* is consistent: a moral ecology in which social justice is rewarded by material prosperity. Both mythical and historical events are reread through this filter, and the deployment of ecological metaphors strengthens the focus on the environment. The idea of the earth having a voice, in particular, is a metaphor that in Biblical contexts has been observed as potentially helping its audience to 'begin relating to the earth as partner and co-creator rather than property' (see Marlow 2009: 124–5 n.22). The Sibylline mode is certain of divine involvement in 'natural' events, although the expression of this faith takes a wide variety of forms. The terminology of destruction in particular, however, leaves plenty of room for ambiguity (are the forms of destruction metaphorical or concrete? Are they punishments or portents?), and the discontinuous, recycled form of the text also provokes questions about the security of the readers in their world. To classify this text as a kind of 'ecocritical' literature is to recognize that one of its driving concerns is attention to the environment, if only as a clue to ethical judgement of each nation. In this focus, the form and content of the *Sibylline Oracles* are mutually reinforcing.

CHAPTER 3
SELF-SUFFICIENCY AS A DIVINE ATTRIBUTE IN GREEK PHILOSOPHY
David Sedley

Only the world is said to be self-sufficient, being the only thing to have within itself everything that it needs. Its nourishment and growth are supplied from within it, as its lesser parts transform into one another.

(Chrysippus, *On Providence* Book 1)[1]

The cucumber is sour? Throw it away. There are brambles in your path? Go round them. That is enough, without your adding 'And why did these things even come to be in the world?' For a physicist will laugh at you, in the way that a carpenter and a shoemaker would laugh at you for complaining that in their workshop you can see shavings and offcuts of their products. On the other hand, they at least have somewhere to throw these, whereas the nature of the universe has nothing outside itself, and the wonder of its craft is that, having set limits upon itself, it takes all the things inside itself that seem to decay, to age and to be useless, and changes them into itself, and that from these very things it recreates other, fresh ones, so that it will never need either substance from outside, or some further place to jettison its refuse. It finds sufficiency in its own space, its own matter, and its own distinctive craft.

(Marcus Aurelius, *Meditations* 8.50)

The second-century CE Roman emperor Marcus here expresses wonder at the world's self-sufficiency, manifested in the way it supplies all its own needs through recycling. Although Marcus certainly writes as a Stoic – witness how closely his words match the preceding quotation from the school's leading exponent Chrysippus – he tilts his forerunner's emphasis in a risky direction when he adds that the universe has in any case nowhere external to dump its waste. The Stoic 'universe' consists of a spherical world surrounded by infinite void[2] – void which might have provided a perfect dumping ground for the waste matter, had the world's immanent divine power so planned things. What a Stoic should be expected to say about self-sufficiency is not that the world has no choice but to recycle, but that its divine causal principle positively chooses that it should do so, self-sufficiency being intrinsically preferable to dependence.

The background to this Stoic doctrine lies in a long philosophical tradition about the self-sufficiency of the divine. Consider the following question. How can human piety involve serving (*therapeuein*) the gods, when the gods are in their very nature self-sufficient beings who lack nothing and need nothing? This puzzle, posed by Socrates in Plato's *Euthyphro* (12e–14c), inspired a variety of solutions, of which the most significant is the one voiced by Plato in his own late work the *Timaeus* (90a–d) and by Aristotle (*Nicomachean Ethics* 10.8, 1179a22–4; *Eudemian Ethics* 8.3, 1249b12–17): the god who does need your service is in reality your own intellect, or inner divinity, which it falls to you to serve or cultivate (*therapeuein*). That gods, properly understood, are fully self-sufficient is a commonplace of ancient philosophical writing, enshrined for example in Aristotle's celebrated dictum (*Politics* 1253a27–9) that someone unable to associate with others or so self-sufficient as to have no need of others would be either a beast or a god.

Our own human self-sufficiency, to the limited extent that we even have it, shows itself in our degree of independence *of other human beings*; and analogously, in Aristotle's eyes, the gods' postulated self-sufficiency requires above all that they have no dependence *on other gods*. Aristotle's gods, that is, not only do not need *us*, more importantly they do not even need *each other*. Hence, Aristotle observes in the *Nicomachean Ethics* (10.8, 1178b8–20), gods can stand in no contractual or other relations of obligation *among themselves*, such as would require them to have and exercise moral virtues. Nor, according to an argument in the *Eudemian Ethics* (7.12, 1244b7–9), do gods either have or need friends. Rather it is natural, in Aristotle's eyes, to attribute to them that same supremely fulfilling activity which we ourselves can perform with the least possible dependence on others (*Eth. Nic.* 10.7, 1177a27–b1), namely intellectual contemplation (*Eth. Nic.* 10.8, 1178b7–22).

Given the theme of this volume, rather than continue to consider gods in general I shall focus on just one kind, namely cosmic gods. From the fifth century BCE onwards our 'world' (*kosmos*) was seen as a unified structure, with the earth at the centre and/or bottom, strata of water and air above or surrounding it, and the whole ensemble bounded by a spherical rotating outer layer, the fiery heaven.[3] Whether this world is the only such structure in the universe (*to pan*, literally 'the all'), or whether there are other worlds, like or unlike ours, always remained open to debate. But naturally the major focus of cosmological speculation concerned the nature and origins of the single world that we ourselves inhabit.

To a 'world' thus understood, our nearest modern equivalent is a habitable planet. For all the obvious differences of perspective that the current heliocentric planetary model imposes, from the point of view of its inhabitants what constitutes the world has changed little. Earth, sea, air and sun are still the four great cosmic players that jointly constitute our environment.

There are however some major differences too. A thesis espoused by three of the ancient philosophical systems – those of Empedocles, Plato, and the Stoics – is that the world either is, or at certain times becomes, a god. And in the work of at least the first two of these thinkers, on whom I shall largely concentrate here, the world's complete self-sufficiency as a perfectly integrated system is prominently presented as a divine attribute. I start with the Sicilian philosopher-poet Empedocles (mid-fifth century BCE).

In Empedocles' cosmic cycle, life first emerges when the great creative force, Love, starts to gain ascendancy over the antithetical force, Strife. At first she can manage no more than to create living beings almost totally lacking in self-sufficiency: isolated arm-creatures, eye-creatures, etc., each with just one specialism, and obviously incapable of sustaining themselves for any length of time. They are nevertheless discrete living beings, and they wander around seeking to team up with others. After a while amalgamations of these beings form complex multi-tasking creatures more or less like those that inhabit the world today; and among them, indeed, the accidentally successful combinations that were sufficiently viable to survive and procreate became the progenitors of today's main species.[4]

Thus in Empedocles' eyes today's creatures, ourselves included, originated roughly speaking as what are now called superorganisms. A superorganism is an organism which consists of organisms. The concept was developed by William Morton Wheeler in the early twentieth century (Wheeler 1911).[5] According to him and others, the extraordinary success of social insects, such as ants, bees and termites, is to be attributed to their genetic programming to act, not each in its own interest, but for the communal advantage of the colony. The colony itself is, in effect, the primary organism, in which the individual insects function like cells or organs with various specialized functions, and which enjoys a degree of self-sufficiency that its individual members entirely lack. The queen, for example, with her reproductive function, is in effect the genitals of the colony.

Not dissimilarly for Empedocles,[6] isolated eye-creatures, foot-creatures etc. could survive only by entering the complex cooperative partnerships in which they now find themselves. It is clear enough that biological progress under Love's influence involves a move towards increased self-sufficiency, which the original 'single-limbed' creatures utterly lacked, but which today's complex creatures possess, albeit still imperfectly.

This snapshot of biological history prepares us for a future state of the world in which, according to Empedocles, it will itself function as a self-sufficient divinity. If we humans, along with oxen, birds and the like are already, loosely speaking, superorganisms, in virtue of having emerged as new species from cooperating teams of simpler organisms, does that constitute the limit of amalgamation, or are we ourselves in turn to be components of a larger superorganism? More specifically, is the world itself the ultimate superorganism? This is the extreme to which the modern conception of the superorganism has been taken by some of its adherents, most notably by James Lovelock, whose Gaia hypothesis treats the earth's entire ecosystem as a single self-regulating organism (Lovelock 1979).

How about Empedocles? When Love finally reaches her zenith and Strife is banished, what the totality then becomes arguably does have some claim to be called a superorganism. This is the god Sphairos.

> There neither is the shining form of the sun discerned, nor the shaggy might of the earth, nor the sea. So dense is the covering of Harmony by which it is held fast: round Sphairos, enjoying his blissful solitude. (B27)

> But equal to himself on all sides, and altogether unbounded, round Sphairos, enjoying his blissful solitude. (B28)

> For there is no pair of branches issuing from his back, no feet, no swift knees, no generative organs. (B29)

After certain very long intervals, Love achieves total dominance and the universe becomes, and thereafter long remains, a single blissful god. Because the universe is spherical, this universe-god is given the name Sphairos, the unique masculine derivative of *sphaira* ('ball' or 'sphere') probably being Empedocles' own coinage. In being grammatically masculine, this god stands apart from the world's two governing powers: the feminine Love (*Philotēs*) and the neuter Strife (*Neikos*). But despite his masculine gender, Sphairos lacks male genitalia (B29). Why so?

Biologically speaking, it is no doubt because he is extremely long-lived and has no need to reproduce. At another level, Sphairos' lack of genitalia is the mark of divine self-sufficiency. A created divinity who is himself the entire universe has no need of our familiar asymmetric external appendages – arms, legs, mouth, etc. Such asymmetries are indispensable for imperfectly self-sufficient beings such as ourselves, condemned to constantly interacting with our environment by travelling through it, ingesting it, excreting into it, copulating with it, warding off the dangers it presents and so on, but a living being identical with the entire universe has no corresponding needs, there simply being nothing external to it. And because Empedocles' universe-god has no need for asymmetries, he remains perfectly symmetrical, in other words, spherical. His lack of male genitalia is in this sense coordinate with the other absences that fragment B29 also lists by way of example: his lack of wings (if that is how the 'pair of branches issuing from his back' is to be understood), feet and knees.

What is this universe-god like? Viewed geographically, he does not have his four constituent elements – earth, water, air and fire – separated into discrete cosmic masses: there is no separate sea, sun or earth, we are told in fragment B27. In this respect, he does not resemble Plato's harmonious universe-god, who although spherical on the outside has a complex cosmic structure on the inside. Hence Empedocles' spherical god has often been inferred to be a perfect *blend* of the four elements; and that, if so, would make him a ball of a single homogeneous stuff. However, several fragments refer to his 'limbs' (B27a, B30, B31), and if he has limbs he is not after all a homogeneous soup, but a complex organism with multiple organic parts.

Thus Empedocles' legacy to Plato is the idea that, at least in ideal conditions, the world is unified into a blissfully happy, organically harmonious, self-sufficient, spherical divinity. Those ideal conditions are not fully satisfied by the present world, in which Strife still retains a foothold. But they do represent its optimal past and future state.

Written a century later, Plato's *Timaeus* sets out a cosmology which draws heavily on this Empedoclean heritage. Plato's world is, likewise, a blissfully happy spherical created god: he calls it 'the self-sufficient and supremely perfect (*teleōtaton*) god' (*Ti.* 68e). The major difference from Empedocles is that the world Plato is describing is not just an ideal past or future world, but the present world too. So successful a creation is the Timaean *kosmos* that it will forever remain in its present ideal condition. It is the handiwork of a supremely good divinity, the Demiurge, to whom there is no contrary force, comparable to Empedocles' Strife, that could ever undo the world's perfection.

Empedocles saw the present world as not yet a unified living being, partly because it is riven by Strife into the four cosmic masses earth, water, air and fire. Plato's speaker Timaeus does not see this present stratification as any kind of disunity, but on the contrary as the only correct way for a world to be unified (*Ti.* 31b–32c): in the process of its original creation its two bounding solids, earth at the centre and fire at the periphery – which were needed in order to provide, respectively, the tangibility and the visibility that are jointly the hallmarks of body – could mathematically speaking be brought into harmony only by the interposition of two mean proportionals; and that is why there are two intermediate bodies, water and air, yielding four strata in all, perfectly harmonized.

Moreover, since Earth is itself a divinity, as are the innumerable fiery beings – Sun, planets, fixed stars – located in the heaven, the Platonic world god who consists of these and other parts enjoys the unification proper to a superorganism. He is an organism that consists of organisms.

Not only is our world 'one' in the sense of being a harmonic whole, it is also 'one' in the sense of being unique. The Demiurge chose to make just one world, primarily for metaphysical reasons (*Ti.* 31a–b) that we can set aside here, but also for the practical motive of using up all the available matter on it, so that nothing with even the potential to harm it should be left outside (*Ti.* 32c–33b). As Plato's spokesman Timaeus presents the outcome of all this reasoning:

> He set it up as one single world, able thanks to its virtue to associate with itself and in no need of another, but satisfied with its own acquaintance and friendship. Thanks to all these things, he created it a happy god. (*Ti.* 34b)

These words encapsulate the primary notion of divine self-sufficiency at play in the *Timaeus*. Like Empedocles' Sphairos, our world neither has nor needs neighbours, being able to be supremely happy exercising its own powers of thought, without any equivalent of a social life. This description makes the world a divine model of the solitary contemplator, and thus a paradigm of the contemplative human life which both Plato and Aristotle held to be more self-sufficient and thus more godlike than any life of purely civic or social virtue could ever be.

Timaeus has a good deal more to say about world design. He never mentions the empirical fact – as it was taken to be by Plato's day – that the visibly rotating outer layer of our cosmos is a spherical shell. Instead he gives us a whole series of theoretical reasons why a good creator was bound in any case to make a spherical world.

These reasons rest largely on the merits of self-sufficiency. The sphere is the one perfectly symmetrical solid, and according to Timaeus by far the most beautiful. What then could have led to the world's being assigned any shape less symmetrical than this one? Asymmetries, for Plato as for Empedocles, are precisely the mark of a *lack* of self-sufficiency. If there had been any matter external to our world-god, Timaeus explains, he might have needed projecting limbs for self-protection, a mouth for ingestion and breathing, and eyes and ears for engaging with his environment. Here are Timaeus' actual words:

On the outside he made the whole of it strictly smooth and circular, for many reasons. (a) It had no need of eyes, since nothing visible was being left outside; or (b) of ears, since nothing audible was being left outside. And (c) there was no air surrounding it that needed to be breathed. Nor (d) did it need any instrument with which to ingest nourishment, and (e) with which to excrete that which had been previously digested. For nothing left it, nor did anything enter it from anywhere, there not even being anything to do so. For it has been skilfully made so as to provide its own waste as its own nourishment, and to accomplish everything both passively and actively within itself and by its own agency. This is because its constructor thought that it would be better for being self-sufficient, rather than needing others. And (f) he did not see any point in attaching hands to it, since it had no need to use these for holding or defending itself against anything. (g) The same goes for feet, and in general for locomotive equipment. (*Ti.* 33b–34a)

Methodologically, much of the fascination of Plato's enterprise lies in the fact that the creation story set out by Timaeus has very little empirical content, but is, rather, a theoretical and speculative reconstruction, or thought experiment, as to how a supremely good creator could be expected to construct a world. For example, it turns out to have been multiply determined that any world created by a good designer should be spherical: not just because, as we have seen, it had no need for prostheses, orifices or other asymmetries, but also because this shape enabled its outer envelope to rotate eternally in its own space and thus think eternal thoughts (*Ti.* 36e–37a). Timaeus' discourse is, from one perspective, a recipe for how a world *should* be designed to function. In other words, what it sketches for us is the *intelligent* way to make a world. That it does also turn out to describe our actual world follows from the additional premise that our world is in fact itself the handiwork of a supremely intelligent and good creator. This premise is assumed from the start of his speech, but never separately justified, since even to entertain the possibility of its not being so would be impious (*Ti.* 29a).

A good creator setting about making a world, according to Timaeus' reasoning, would be bound to set it up so as to be everlasting. This does not amount to making it altogether indestructible – since whatever can be constructed can also be dismantled – but rather to his joining it with bonds so tight that no one but he himself could undo them. (Think of tying a knot which only someone as strong as you could untie, and then imagine the equivalent for the most powerful being in the universe.) The further, crucial point in Timaeus' reasoning is that a benevolent creator god, although *capable* of dismantling or shutting down his creation, would have absolutely no motive for ever doing so, especially given that it is the best possible. From these reflections one thing that plainly emerges is Plato's recognition that a world's ability to survive into the indefinite future is among its primary merits. For his purposes, this everlasting durability is necessary if the world-god is to possess that archetypically divine attribute, immortality. For us, however, the bells that it rings will have more to do with sustainability.

Earlier we encountered Timaeus' picture of the world-god's entirely self-contained recycling of waste – recognizably the original that inspired the passage of Marcus

Aurelius at the head of this chapter. Marcus, although a Stoic, suspended judgement regarding the majority Stoic thesis that the world has only a finite duration,[7] ending in a periodic conflagration (*ekpyrōsis*) after which it is recreated. This same doctrine had been rejected outright by some previous Stoics, notably Panaetius (late second century BCE). We may conjecture that a principal reason for such Stoic doubts lay precisely in the legacy of our Timaean passage with its description of the world-god's perfectly self-contained recycling system, which for Plato is the key to that god's ability to retain his optimal structure and functioning in perpetuity.

It is important to add that the structure which according to Timaeus is in this way perpetually maintained is not the bare framework of a world, constituted just by the main cosmic habitats – land, sea, air and the celestial fires. It extends all the way down to the maintenance of the world's resident species. For the world's completeness as an all-embracing living being or animal is said by Timaeus to depend on its containing – much as the genus animal itself does – all the animal kinds (*Ti*. 41b–d). Plato believes, more literally than most of us do, that it takes all sorts to make a world: the world-animal would not be a complete world if it did not contain every single animal species. Thus Plato's conception of a world designed to last forever *includes* its perpetuation of maximum biodiversity. As for humans, although Plato considers them the best mortal species, there is not so much as a hint that the world has been created for their sake.

For our part, we might want to bring Plato's characterization of what it would take to constitute a fully self-sufficient cosmic divinity into dialogue with our own world's increasingly pressing claims on self-sufficiency. Three main points may claim our attention. First, if the Platonic cosmos is maximally isolated, being quite literally the only world in the universe, this reminds us that our planet is sufficiently isolated to be, now and in perpetuity, the only world we can call upon to supply our needs. According to one calculation, if our world's entire population were to acquire the European lifestyle we ourselves enjoy, it would take three Earth-like planets to supply its needs – or four, if the lifestyle in question were American.[8] Plato reminds us that we have only the one.

Second, he reminds us that our world, like the intelligently run world he describes, can aspire to long-term survival only by endlessly recycling its own internal resources.

Third, his cosmic theology specifies that a well-made world should be *biologically* everlasting, with reference to perpetual maintenance not only of the cosmic superorganism itself, but also of full biodiversity among its resident species.

With or without the theological undercurrents, these look like materials for a fruitful dialogue between Plato and ourselves on that pressingly urgent question – how to treat a self-sustaining biosphere with the reverence that is its due.

CHAPTER 4
A LIGHTER SHADE OF GREEN: STOIC GODS AND ENVIRONMENTAL VIRTUE ETHICS
Christoph Jedan

> It would be a long story to tell of the services rendered by mules and asses, which were undoubtedly created for the use of men. As for the pig, it can only furnish food; indeed Chrysippus actually says that its soul was given it to serve as salt and keep it from putrefaction; and because this animal was fitted for the food of man, nature made it the most prolific of all her offspring.
>
> Cicero, *On the Nature of the Gods* 2.159–60; trans. Rackham 1933

> Look how many things extravagance, the plunderer of land and sea, has mixed together so that they fit through a single throat.
>
> Seneca, *Letters* 95.19[1]

Any attempt to assess Stoic thinking in light of our current ecological crisis pulls us in opposing directions. On the one hand, Stoicism has been blamed for the anthropocentric attitude that underpins modern technological civilization and its destructive attitude towards nature; on the other, Deep Ecology's biocentric thinkers regularly acknowledge the neo-Stoic Spinoza as a major influence (e.g. Sessions 1977). Some would argue that we cannot cast the Stoics as precursors of today's Deep Ecology – the instrumentalist view of a species that is evident in the first quote above renders futile any such attempt – whilst others would maintain that the ancient Stoics certainly did not license unbridled exploitation of nature, as the second quotation above confirms. It is probably only safe to say, as some modern commentators have done, that the Stoics occupy some middle ground between biocentric egalitarianism and unbridled dominion over nature. This moderate stance, however, need not rule out a constructive contribution to environmental ethics (e.g. Holland 1997; Dragona-Monachou 2010).

In the present chapter, I will focus on an aspect of Stoic philosophy that has by and large been backgrounded in the debate on Stoicism and the environment: the role of Stoic theology. My contention is that we need to emphasize Stoic theology to understand the full richness of Stoic thinking about the environment. I will argue that there is no

---

My thanks go to the organisers and participants of the conference 'Greening the Gods' in Cambridge 2014 for their valuable questions and suggestions, and to Myrto Dragona-Monachou for sending me the Greek manuscript of her currently out-of-print paper 'Universal Nature and Man in Marcus Aurelius' Stoicism', published in 2010. Work on this chapter was supported by a fellowship at the Royal Netherlands Academy for Arts and Sciences' Institute for Advanced Study (NIAS) in 2016/2017.

Stoic theology in the singular. What we hear frequently referred to as 'Stoic theology' is in reality the interaction of *three* different theologies: Stoic pantheism, polytheism and henotheism. These three theologies are not, strictly speaking, fully compatible. However, in Stoic thinking about the environment they work together, each enriching a delicate balance between, on the one hand, defence of the special status of human beings due to their possession of reason, and on the other hand, refusal to endorse certain forms of hubristic anthropocentric thinking that could give rise to later charges of an unbridled human dominion over nature.

The chapter consists of a short survey of important trends in the literature, setting in context the specific focus of the subsequent sections. They discuss, in turn, Stoic thinking about the gods and Stoic axiology and virtue ethics. The chapter concludes with the application of an important conceptual differentiation in the field of religious studies to Stoic thinking about the environment.

Trends in the literature

The literature on Stoicism and environmental ethics can be characterized by three themes: the sidelining of Stoic theology; critiques of Stoic anthropocentrism; and the promotion of pro-environmental attitudes through Stoic virtues.

Sidelining of Stoic theology

When Alan J. Holland, in his highly stimulating treatment of Stoicism and the environment, refers to Stoic arguments in Cicero's *On the Nature of the Gods*, he writes:

> If we now peruse Cicero's text … we find that the claim that the world and all it contains was made for the sake of [gods and] men, although it is affirmed by Balbus (*Nat. D.* 2.53), is nowhere attributed to Chrysippus. (1997: 154)

To be sure, Cicero did not bracket off the gods; this is a modern author unwittingly exposing the general trend of the discussions on Stoicism and environmental ethics. In these discussions, the role of Stoic theology has regularly been downplayed. And where Stoic theology *is* highlighted, this is done in the context of a *critique* of ancient Stoicism. Pertinent examples are Jim Cheney's 'The Neo-Stoicism of Radical Environmentalism' (Cheney 1989) and Evangelos Protopapadakis' 'The Stoic Notion of Cosmic Sympathy in Contemporary Environmental Ethics' (Protopapadakis 2012). Cheney applies a 'subtextual analysis' to Stoicism:

> The reading of a subtext is the discernment of a hidden agenda underlying the text itself, an agenda which stands in complex relationships to the agents participating in the construction of the text. (1989: 294)

This subtextual analysis allows him to unveil Stoicism's hidden agenda, which comes down to a deeply problematic avoidance of genuinely recognizing and embracing the 'other' (1989: 302):

> The (or a) subtext of Stoic philosophy is the desire for embodiment in some greater whole coupled with a refusal to give up the hegemony that went with Athenian supremacy. The result was alienation from the real world, the Alexandrian world of difference, and the creation of an *abstract* embedment which provided for an analogue of the security of the polis.

The very same subtext, Cheney claims, is at work in the 'Ecosophical' outlook underpinning Deep Ecology (1989: 302):

> Subtextually, the central operative idea at work … is the idea of *containment*, containment of the other, of difference, rather than genuine *recognition* of the other, genuine acknowledgment and embracing of the other.

Similarly, Protopapadakis claims that there are affinities between 'Ecosophy T' as formulated by Arne Naess and Posidonius' notion of cosmic sympathy, only to conclude that cosmic sympathy radically limits moral responsibility and thus undermines the ethical credentials of the theory: 'It seems that Naess's environmental ethics, while surely environmental, is barely an ethics' (2012: 302).[2]

Both the sidelining and the critique of Stoic religion can readily be understood, though not justified, in light of the dialectical context at the onset of the debate on Stoicism and environmental ethics. In 1967, the journal *Science* published Lynn White Jr's article 'The Historical Roots of Our Ecologic Crisis'.[3] White claims that Christianity 'bears a huge burden of guilt' for our scientific-technological civilization, with its negative impact on the environment. According to White, 'Christianity is the most anthropocentric religion the world has seen', exhibiting a stark dualism between man and nature. Firing salvos in quite different directions, White blames *inter alia* the invention of a new plough in seventh-century northern Europe; Latin Western Christianity with its voluntarist tradition; a wave of democratization that had led to a fusion of science and technology in the nineteenth century; and the failure of the West to come up with a new set of values in a post-Christian age. White at the same time praises the Beatnik generation's interest in Zen Buddhism, recommends Saint Francis as a heretical thinker with whom to correct 'orthodox Christian arrogance', and prescribes 'Since the roots of our troubles are so largely religious, the remedy must also be essentially religious, whether we call it that or not' (1967: 1207).[4]

When John Passmore published in 1974 the seminal text for what we with hindsight identify as the nascent discipline of philosophical environmental ethics, he took issue with White's undifferentiated attack on Western culture and mainstream Christianity. Passmore was careful not to write off the Western intellectual tradition lock, stock and barrel, and not to throw out critical reasoning and intelligibility. Christianity, argues Passmore in a chapter evocatively entitled 'Man as Despot', is no uniform movement.

The Hebrew heritage of the Old Testament is free from the idea of man's unbridled mastery over nature. Ultimately, Passmore claims, the roots for Christianity's rampant anthropocentrism lie in Greek philosophy, and in Stoicism in particular. He writes:

> Cicero's Balbus tells us that 'we alone have the power of controlling the most violent of nature's offspring, the sea and the winds, thanks to the service of navigation … Likewise the entire command of the commodities produced on land is vested in mankind … The rivers and the lakes are ours … we give fertility to the soil by irrigating it, we confine the rivers and strengthen or divert their courses … By means of our hands we try to create as it were a second Nature within the world of Nature.' Balbus, then, for all the technological limitations of Greece and Rome, already saw man as a demi-god, constructing with his hands a new nature. (1974: 17–18)

According to Passmore, it was due to their taking over Stoic ideas that Christian authors such as Paul and Origen neglected the far more environmentally friendly Old Testament heritage. This way, the Stoics are held responsible for detrimental Western attitudes toward nature as long as Christianity reigned supreme. What is more, Passmore holds Stoicism also responsible for perpetuating the ideal of human mastery over nature in Western culture after the decline of religious adherence in the West, and he also claims that Stoicism was instrumental in exporting the message of unbridled dominion to other cultures (through Descartes, who inherited the Stoic ingredients of Christianity):

> The ideal of mastery could thus persist in Europe even when the Bible had lost much of its old authority; it could be transferred to countries like China where the Bible had no weight. (1974: 21)

Whether in a Christian or in a secular age, Passmore holds Stoicism responsible for teaching the ideal of human mastery over nature. His picture of Stoicism's universally detrimental role depends on his view of Stoicism as an essentially non-religious movement that could be adapted by different cultures and ages.

Critiques of Stoic anthropocentrism

The charge of anthropocentrism has received much attention in the literature since the writings of White and Passmore, but the debate has evolved considerably since those times of undifferentiated accusations. It is probably best to use a definition of anthropocentrism to identify how the debate has progressed. Anthropocentrism should be understood as 'teleological anthropocentrism', i.e. 'the belief that the whole of creation exists for the sake of humanity' (Attfield 2014: 30). The position can be unpacked as arguing for three theses: (a) an 'instrumentalist view of nature', (b) a 'dualistic representation of nature and of human rationality' and (c) a 'self-divided conception of human nature …, a conception in which people locate the essence of humanity in rationality, and contrast this with their physical and emotional nature'

(Attfield 2014: 31, see also Attfield's chapter in this volume, pp. 75–89). There is no reasonable doubt that the Stoics fit some interpretation of thesis (a), since they tended to view nature as created for the sake of *inter alia* human beings. It is a different question, however, whether or not the Stoics would have deduced from thesis (a) an unbridled *dominion*, i.e. the right of human beings to do with nature as they please. Since Passmore, the tendency has been to emphasize elements of Stoicism that go against ideas of an unbridled dominion. As to theses (b) and (c), there is no straightforward sense in which they apply to Stoicism. What the Stoics value in nature is reason. Reason, the expression of divine creative power, permeates the whole cosmos. It is present in different degrees everywhere, most of all in the reasonable entities, gods and human beings. Thus nature is made for them, and human beings should strive to become as reasonable as possible. Since reason is the constituent of the whole of nature, including the whole of human nature, there is no space for simplistic differentiations between human rationality and human emotions, or between reason and matter: human beings are part of nature, however much they are invited to see other creatures as made for their benefit. Considering this, the easy charge of anthropocentrism does not make sense, and some observers have rightly suggested that 'logocentrism' would be a far better label for Stoicism (e.g. Stephens 1994; Castelo 1996).

Pro-environmental attitudes and Stoic virtues

Recent commentators have highlighted the positive role Stoicism can play in our attitude towards the environment, either by exonerating individual Stoics from the blame attributed to Stoicism at large or by pointing out where Stoicism can inculcate character traits that might be helpful in facing today's ecological crisis.[5] In this vein, Alan J. Holland has pointed to Stoic fortitude:

> We have seen reason to believe that the very fortitude that the term 'stoicism' now conjures up could be put to the service of environmentalism if it makes us rethink our attitudes to pain and happiness. (1997: 165)

Christopher Gill, seemingly without being aware of the earlier literature, has made similar remarks in a recent post on the blog 'Stoicism Today':

> Although it is sometimes suggested that this crisis can be somehow managed in a relatively effortless way by technological progress, this seems to me largely wishful thinking. ... The Stoic view of ethical development as, on the one hand, demanding in its aspirations, and, on the other, requiring us to work out for ourselves the specific actions that virtue involves, thus offers a good general framework for an effective response to the environmental crisis. (2016)

These observations tie in with the recent emergence of environmental virtue ethics, focusing on 'the proper dispositions or character traits for human beings to have regarding their interactions and relationships with the environment' (Sandler and Cafaro 2005: 3; see also Treanor 2014).

Looking back on the trends in the literature, I argue that what is needed to reinvigorate debate today is a reconsideration of the contribution of the Stoics to an environmental ethics in light of their *theology*, in particular how their thinking about the gods informs the central role of reason in their axiology. Following in the footsteps of thinkers such as Holland, I argue that this re-appreciation of Stoicism's ecological potential should focus on Stoic virtue ethics. In what follows I will draw out these lines, dealing first with Stoic thinking about the gods, then with Stoic axiology and virtue ethics.[6]

Stoic theologies

In reconsidering the central role of theology for Stoic environmental thinking, we should avoid the mistake of pressing it into a rigid mould. In fact, as Keimpe Algra has stated, the Stoic conception of God was 'rather fluid' (2003: 169). The Stoics tried to amalgamate in their theology a wide range of counteracting cultural currents, quite possibly without realizing the full extent of their diversity (Jedan 2009: 31–48, 2010: 25–44). I argue that the Stoics came up not with one, but with *three* theologies that were up to a point at odds with each other: they formulated a form of pantheism, according to which the world as a whole is divine. In addition, they also appropriated a polytheistic rhetoric: they spoke of different more or less human-like godheads having reason and intentions, living together under the rule of law and justice. Finally, they participated in, and contributed to, the henotheistic tendencies of their time: attempts to cut down on the number of divine entities by interpreting them as manifestations of a single underlying divine principle. Stoic speculation about a fiery substance that permeates and unifies the cosmos should be interpreted in this light.

In the second book of Cicero's *On the Nature of the Gods*, our best source for the details of Stoic theology, we find all three theologies in close proximity, without any indication that they were felt to be incompatible. Let me show this by highlighting three passages. In our first passage Balbus, Cicero's Stoic spokesman, presents an argument for the existence of the gods that focuses on a *divine element* that permeates nature:

> There is therefore an element that holds the whole world together and preserves it, and this [is] an element possessed of sensation and reason; since every natural object that is not a homogeneous and simple substance but a complex and composite one must contain within it some ruling principle, for example in man the intelligence, in the lower animals something resembling intelligence that is the source of appetition. With trees and plants the ruling principle is believed to be located in the roots. I use the term 'ruling principle' as the equivalent of the Greek *hēgemonikon*, meaning that part of anything which must and ought to have supremacy in a thing of that sort. Thus it follows that the element which contains

the ruling principle of the whole of nature must also be the most excellent of all things and the most deserving of authority and sovereignty over all things. (*Nat. D.* 2.29, trans. Rackham 1933)

This is an attempt to find a corporeal vehicle for the divine. The divine element permeates nature, it is everywhere, but in different degrees. At its most basic it 'holds the whole world together and preserves it', and is more manifest in the 'ruling principle' of animate beings, coming into its own in the reasonable capacity of human beings. The passage balances two attitudes towards nature pulling in different directions: the first is to emphasize how much all beings are interconnected because they all share in the divine corporeal element. The second is to stress that the divine element is present in human beings in a special degree. Following the latter line of thought in isolation, one could also argue for a dualism: human beings have a quasi-divine status that separates them from the rest of nature.

We find the same balance in our second passage. Here, Balbus argues for *pantheism*, i.e. a divine status of the world as a whole. On the one hand, again, pantheism could be used to affirm the interconnectedness of all beings as parts of a greater, divine whole; on the other hand, the divine nature of the world is argued for by use of teleological arguments and analogies that emphasize the special position of human beings:

For nor is there anything else besides the world which has nothing missing, and which is equipped from every point of view, perfect, and complete in all its measures and parts. As Chrysippus cleverly put it, just as the shield-cover was made for the sake of the shield and the sheath for the sake of the sword, so too with the exception of the world everything else was made for the sake of other things: for example, the crops and fruits which the earth brings forth were made for the sake of animals, and the animals which it brings forth were made for the sake of men (the horse for transport, the ox for ploughing, the dog for hunting and guarding). Man himself has come to be in order to contemplate and imitate the world, being by no means perfect, but a tiny constituent of that which is perfect. But the world, since it embraces everything and there is nothing which is not included in it, is perfect from every point of view. How then can it lack that which is best? But nothing is better than intellect and reason. Therefore the world cannot lack these. Therefore Chrysippus did well to prove by appeal to analogies that all things are better in perfect and mature specimens – for instance, in horse than in foal, in dog than in pup, in man than in child. Likewise, he argued, that which is the best thing in the whole world should be found in something which is perfect and complete. But nothing is more perfect than the world, and nothing better than virtue. Therefore virtue is intrinsic to the world. Indeed, man's nature is not perfect, yet virtue is achieved in man. Then how much more easily in the world! Therefore there is virtue in the world. Therefore the world is wise, and hence is god. (Cicero *Nat. D.* 2.37–39, trans. LS 54H)

It is important to note the strong axiological commitment expressed in the passage: what the Stoics value above all is reason and virtue. Human beings enjoy a special status because they possess reason and can acquire virtue. The special status is thus defended on the basis of a criterion, it is not an instance of 'speciesism', i.e. an unjustified preference for one's own species. Recent commentators have therefore been absolutely right in characterizing Stoicism as a 'logocentric' instead of an 'anthropocentric' system of thought.

Further evidence for the reason-centeredness of Stoic philosophy can be found in our third passage, which presents a *polytheistic theology*. The world is now presented as the dwelling place of gods and men. Both species have reason and virtue or can (in the case of human beings) acquire virtue:

> In the first place the world itself was created for the sake of gods and men, and the things that it contains were provided and contrived for the enjoyment of men. For the world is as it were the common dwelling-place of gods and men, or the city that belongs to both; for they alone have the use of reason and live by justice and by law. (*Nat. D.* 2.154–5, trans. Rackham 1933)

Again, the passage balances two attitudes: on the one hand, human beings enjoy a special status, but on the other hand, that special status is not described in such a way as to offer a motif for a speciesist hubris. In fact, one Stoic argument for the existence of gods uses the unacceptability of human speciesist hubris as a premise. In *Nat. D.* 2.16, Chrysippus is quoted as arguing that it would be 'stupid arrogance' to suppose that there would be nothing better than mankind in the cosmos. So Stoic polytheism too helps to maintain a very delicate balance: human beings are special creatures, due to their possession of reason, but they are certainly not the only species in the world that matters.[7]

To recapitulate, the available evidence of the three Stoic theologies supports the interpretation that the Stoics attempt to preserve a delicate balance between a special position of human beings and their embeddedness in a larger whole. Theirs is a 'moderate anthropocentrism' or, rather, a moderate logocentrism.[8] But how can the interpretation of the Stoics as *moderate* logocentrists hold in light of Chrysippus' seemingly dismissive remark about pigs quoted at the beginning of this chapter? The fact that the teaching in question is also mentioned elsewhere makes it impossible to explain the quotation away as misunderstanding. We need to interpret the quotation, however, in light of its most likely dialectical context.[9]

If we can take our cue from Cicero's presentation in *On the Nature of the Gods* book 2, the quotation seems to be taken from Chrysippus' arguments for the gods' providential care for mankind. Providential care is linked by the Stoics to a teleological structure of the world that leaves human beings well provided for. In this argumentative context, Chrysippus could have easily referred to animals whose usefulness for human beings was generally admitted, but he also would have had to answer questions about animals whose usefulness was a matter of doubt or even ridicule. We see a reflection of such an exchange in Plutarch's *On Stoic Self-Contradictions* (1044B–F), polemically recontextualized by Plutarch as a discussion on moderation or even austerity as a civic virtue. Plutarch parades a Chrysippus who in his books on natural philosophy is clearly bent on showing that even bed bugs and mice are useful since the former help us to wake up and the latter stimulate us to be tidy. Chrysippus goes so far in finding purpose in animals as to argue that the peacock was made for the sake of his beauty, a position Plutarch critiques because he finds it to contradict Chrysippus' condemnation of luxury in contemporary societies. Now, I suggest that mentions of the pig, an animal widely held in contempt in antiquity, would have served a similar argumentative purpose as mentions of bed bugs and mice, challenging Chrysippus' attempt to find purposeful providence in nature. Chrysippus very likely had to counter an argument to the effect that pigs do not fulfil a genuine purpose in the greater providential scheme of things: a comparable nutritional value could have been provided by other means. Only in such a context can we make sense of Chrysippus' combining mentions of exceptional nutritional value, food conservation and the animal's soul. So, if I am right, Chrysippus' remark is an argument that *defends* the rightful place of a species in the larger framework of divine providence. While the remark is certainly not up to today's standards of environmental correctness, it represents a considerable step up from the unmitigated contempt for pigs that fuelled the argument which Chrysippus attempted to refute.

Stoic axiology and virtue ethics

What the Stoics value in the world is its reasonable, teleological structure. The faith of the Stoics in the world's reasonable structure is such that it withstands, and is harmonized with, their recognition of nature's destructive side. The Stoics viewed the history of the cosmos as cyclical. In recurrent conflagrations, the divine element consumes the other elements and remains alone until a new world cycle begins. The souls of human beings, even the gods of polytheism, are said to perish during conflagration.[10] The narrative of cyclical destruction and reconstruction allowed the Stoics to interpret particular experiences of loss against the background of an all-encompassing eschatology. Again, the three Stoic theologies intertwine (radical change of the whole world as divine – continuity of the divine element – perishability of polytheistic godheads) in making destruction part and parcel of the world cycle. This has momentous consequences: ultimately the doctrine of cyclical conflagration denies that radical change might destroy something valuable for good.[11]

The Stoics react to the impermanence of the world, which they declared to be part of its reasonable structure, by radically divesting of value everything apart from the reasonable structure itself. The famous Stoic distinction between what is good or bad and what is indifferent ought to be understood in this vein:

> They [the Stoics] say that some existing things are good, others are bad, and others are neither of these. The virtues – prudence, justice, courage, moderation and the rest – are good. The opposites of these – foolishness, injustice and the rest – are bad.

> Everything which neither does benefit nor harms is neither of these: for instance, life, health, pleasure, beauty, strength, wealth, reputation, noble birth, and their opposites, death, disease, pain, ugliness, weakness, poverty, low repute, ignoble birth and the like. (Diogenes Laertius 7.101–2, trans. LS 58A [part])

The Stoics identified a core sense of good, in which only virtue and its opposite, vice, are good or bad respectively. The things ordinarily held to be (non-moral) goods or evils are strictly speaking 'indifferent'. They are not constituents of the good life (*eudaimonia*), since the good life is possible without such (non-moral) 'goods' and with such (non-moral) 'evils'. The demotion of non-moral goods and evils as 'indifferent' does not leave our decision-making completely rudderless, however, since many indifferents are capable of activating 'impulse' or 'repulsion'.[12] This is something that is instilled by observation of nature and in this sense those things that activate our impulse are called 'preferred indifferents' or 'in accordance with nature', whereas those things that activate our repulsion are called 'dispreferred indifferents' or 'contrary to nature'.[13] However, whilst the differentiation between preferred and dispreferred indifferents offers some guidance, the primacy of moral value means that any such guidance can only be provisional. Considerations of virtue override any impulses or repulsions.[14]

Thus, another famous slogan of the Stoics, to 'live in agreement with nature' should not be misconstrued as a theory of goods emphasizing the independent value of every living being and putting forward a conservationist agenda. Stoicism teaches no such thing. The identification of 'living in agreement with nature' with 'living in accordance with virtue' brings home, again, the primacy of virtue. Nature teaches virtue as the perfection of reason in human beings and thus the completion of the human task of appreciating the world's reasonable structure and bringing it to bear on one's life.

> Zeno in his book *On the Nature of Man* was the first to say that living in agreement with nature is the end, which is living in accordance with virtue. For nature leads us towards virtue. So too Cleanthes in his book *On Pleasure*, and Posidionius and Hecato in their books *On Ends*. Further, living in accordance with virtue is equivalent to living in accordance with experience of what happens by nature, as Chrysippus says in *On Ends* book 1: for our own natures are parts of the nature of the whole. Therefore, living in agreement with nature comes to be the end, which is in accordance with the nature of oneself and that of the whole, engaging in no activity wont to be forbidden by the universal law, which is the right reason pervading everything and identical to Zeus, who is this director of the administration of existing things. (Diogenes Laertius 7.87–8 = LS 63C [part])

All this means that Stoic ethics represents a radical type of virtue ethics that makes rightness dependent on virtue.[15] Commentators who are convinced that virtue ethics does not get the order of dependency right – one ought to have an independent standard of value and derive virtues from that standard as character traits that promote the values thus instituted – are bound to be disappointed with Stoic ethics, the more so since virtue ethics retains the focus on the human agent, which some claim denies the intrinsic value of nature.[16] In the present context I cannot dive into a full discussion of meta-ethics and its relevance for environmental ethics. Suffice it to say that the virtue-ethical focus on the human agent may actually be a good thing, since it has decidedly *practical* and *motivational* advantages over consequentialist theories, which focus on bringing about the best outcomes. In a situation of negligible individual impact ('one plastic bag more or less does not matter'), consequentialist accounts will have a hard time creating a sense of urgency. Virtue ethics, and Stoic virtue ethics in particular, provides an excellent antidote to the threat of the negligibility of individual actors. The Stoics offer a clear-cut distinction between what is done by an actor and how something is done. With their heavy focus on the 'how' side of human action, the Stoics bring home the message that virtue is supremely important. Our achievements in terms of non-moral goods and the visibility of one's impact on the world at large do not matter so much. A Stoic would feel motivated to do the right thing by the environment whether or not their own actions bring about results that are immediately visible. Individual negligibility is no issue for the Stoics.[17]

On the basis of the discussion so far we are equipped to appreciate the specific profile of a Stoic contribution to environmental ethics. The Stoics, who were convinced that health, pleasure, beauty and even life were (albeit preferred) 'indifferents', could not conceivably argue that the continued existence of any species presents an absolute intrinsic value which cannot, under any circumstances, be trumped. What they could focus on, instead, was the *human* traits of character that are expressed in exploitative or protective attitudes towards the environment, and this is precisely the strategy that we see at work in the second quotation with which I opened this chapter:

> Look how many things extravagance, the plunderer of land and sea, has mixed together so that they fit through a single throat. (Seneca, *Letters* 95.19)

Seneca delivers in his ninety-fifth letter a clear message of moderation or even frugality, against the extravagance (*luxuria*) he sees everywhere around him. As we must expect in the Stoic virtue-ethical framework, Seneca's starting point is the value or disvalue of certain traits of character, not an intrinsic value of the environment.[18] The immediate focus of the letter is the prescriptive part of philosophy, more particularly the correct relationship between fundamental doctrines (*decreta*) and more specific practical rules (*praecepta*). In this context he raises the question of why the prescriptions of philosophy seem so powerless. Seneca blames rampant 'moral perversion' (*morum perversitas*: 95.34), which to him is manifest in extravagant food consumption (ever more complex combinations of ingredients, overeating, intoxication with alcohol), sexual debauchery and cruelty against fellow human beings. He declares:

> Against such a moral deterioration nothing should be left undone (*adversus tantam morum eversionem omnia conanda sunt*). (*Letters* 95.29)

To repeat, the fact that Seneca's critique focuses on human character instead of an intrinsic value of every form of life might be a disappointment to some, but once we are prepared to appreciate what a specific form of ethics can and cannot do, Stoic virtue ethics becomes a useful voice. Recently, Peter Wenz has suggested that the traditional virtues of 'frugality, appreciation, temperance, self-development, dedication, benevolence, generosity, empathy, and justice' counter consumerism (2005). Since consumerism relies on traditional vices such as greed, luxury and pride that undermine human flourishing, there are two complementary justifications of the traditional virtues:

> [N]onanthropocentric environmentalists have reasons to favour traditional virtues because their exercise tends to protect the non-human environment. Anthropocentrists have reason to support the same virtues because their exercise promotes human flourishing. (Wenz 2005: 197)

I side with Wenz. Even nonanthropocentrists should recognize the positive contribution of the virtues in terms of environmental protection and human flourishing. Perhaps it is no coincidence that Wenz's list of environmentally friendly virtues appears remarkably close to Stoic virtue tables which list *inter alia* moderation, justice, courage, self-control, resourcefulness and kindness.[19] With his polemic against extravagance, the 'plunderer of land and sea', Seneca represents mainstream Stoic thinking.[20] Far from being the instigator of our environmental crisis, Stoicism might offer interesting resources to fight exploitative attitudes towards nature.

Conclusion

In his 2010 book *Dark Green Religion*, Bron Taylor argues that many environmental movements in the United States today can be described as new, emerging religions of nature, a trend for which he identifies Spinoza and Rousseau as 'early exemplars'. His analysis is premised on a conceptual differentiation that has gained quite a bit of traction in the field of religious studies:

> It is important to distinguish between green religion (which posits that environmentally friendly behavior is a religious obligation) and dark green religion (in which nature is sacred, has intrinsic value, and is therefore due *reverent care*). (Taylor 2010: 10, my italics)

It should be clear from our discussion so far that the ancient Stoics do not qualify as 'dark green' thinkers in Taylor's sense. Certainly, the three Stoic theologies allowed for a conceptualization of *the whole of nature* as sacred and in all likelihood the Stoics could have agreed to attribute intrinsic value to it. The Stoics could thus see nature as a unity that human beings must revere. Taylor and the Stoics part company, however, in respect of an attitude of *reverent care*. There is no way in which the Stoics could fathom nature as being in need of our care.

I want to argue that the reason for this inability is not, or at any rate not primarily, as Hans Jonas (1979: 20) has suggested, the natural tendency to conceive of the whole world as invulnerable, a presupposition that became questionable only in recent times, with the growing realization of the destructive power of technological civilization. Whilst the Stoics too might have shared that 'natural tendency', I want to suggest that there is a deeper difference between Stoicism and Taylor's dark green religion and that this difference turns on questions of theology. Ultimately, Taylor's concept of dark green religion appears to derive from a Christian worldview that would have been alien to the ancient Stoics.

For nature as a whole to be conceptualized as the object of our reverent care, as Taylor suggests, we need to see nature not only as a godhead – this is achieved in Deep Ecology through the Gaia hypothesis and is also in line with ancient Stoicism – but also as a vulnerable, mortal god. Human beings have to be as it were the eyes and hands of this godhead; our contribution matters on a cosmic scale. Those lines appear to be central ingredients of Christian theology. Moreover, the sense of urgency that is needed to get any form of activism going is dependent on a Christian view of history as linear, from creation to redemption. Secularized Christian worldviews deleted the end stage of redemption from this picture, but they kept the linear view of history. Only against the background of history as linear can the loss of biodiversity, the degradation of the natural environment and the loss of prospects for human life be seen as a tragedy that must create a sense of urgency. If I am right in suggesting all this, there are probably far closer connections between Christian theology and recent ecological thinking than many Deep Ecologists acknowledge.

The Stoics, by contrast, regarded nature *inter alia* as an encompassing godhead involved in a process of cyclical destruction and regeneration. Human beings ought to appreciate and venerate that process, and learn to regulate their behaviour according to standards derived from the close observation of the cosmic cycle. Against this background, the depletion of natural resources, the loss of biodiversity, the perishing of human beings and so forth are part of a destructive and regenerative cosmic process. What is lost can find a repeat in the next world cycle. Loss does not register as a tragedy.

This does not mean that Stoicism precludes vigorous action against degradation of the biosphere. On the contrary, the example of Seneca suggests that there are fair prospects of drawing inspiration from Stoicism for an environmental virtue ethics. However, the sensibilities will differ considerably from the dark green religions in Taylor's conceptual framework. Stoicism represents a lighter shade of green.

CHAPTER 5
COSMIC BEAUTY IN STOICISM: A FOUNDATION FOR AN ENVIRONMENTAL ETHIC AS LOVE OF THE OTHER?

Jula Wildberger

This chapter offers an interpretation of the Stoic concept of beauty that might serve as a starting point for an eudaemonist environmental ethics. It is based on the following two assumptions. First, humans are not only rational beings but also ones who find their ultimate value in loving and caring for what is other than themselves. Second, this motivational orientation aligns human beings well with the world around them, so that they are capable of 'expressing' that trait successfully and flourish when doing so.[1] Even though it developed out of an historical interest in ancient Stoicism and will discuss relevant sources for that school of thought, the interest of this chapter is systematic. I do not offer a faithful reconstruction of a theory actually held by any individual ancient Stoic. Rather, I wish to acknowledge my debt while developing an idea that is present in the transmitted discourse and can be made productive within a Stoic theoretical framework. My claim is that the beauty which Stoics attributed to the cosmos can be understood as a consequence and expression of the active principle God's sociability and as a term for the implementation of a peculiar form of cosmic individualist utilitarianism. I will first present my reading of Stoic sociability and show how the Stoics connected this disposition in humans with sociability as a divine and thus also cosmic feature. Then I will analyse the Stoic concept of beauty and indicate its role in accounts of Providence and also show how this concept overcomes the dilemma faced by individualistic and holistic approaches in modern environmental ethics. I will conclude with some suggestions as to how the Stoic concept of beauty might support a modern environmental ethic that, unlike ancient Stoicism, does not presuppose theism and a providentially ordered world.

Sociability

The ancient Stoics are theists, at least to the extent that they assume an individual divine creator of this world, whom they name God and to whom they ascribe intentions and actions.[2] These assumptions give rise to a question: Why would God make a world? God

---

I would like to thank the Max-Weber-Kolleg, Erfurt, and the colleagues there for the time, funding and inspiration that come to fruition in this chapter.

by himself is an eternal, imperishable perfect being, perfectly rational and self-sufficient, and not in need of anything. So why would such a being make a world? The Stoic divinity has to invest a lot in order to make the cosmos. It is not just an affair of intelligent design and setting a cosmic clockwork machine in motion. God is not an omnipotent transcendent being that can bring about whatever world he likes in a moment of instantaneous creation. Stoics believed that causation requires physical contact between one body, the cause, and the body on which the cause has its effect. Accordingly, God, whom they also called the 'active principle' (*to poinoun*) and the Cause of everything, has to physically mix with inert, unqualified Matter (*hule* or *ousia*), which they called the 'passive principle' (*to paskhon*). Choosing to act upon Matter in this way, God embraces all kinds of limitations to his agency.[3] By continuous movement and activity everywhere in the cosmic God-Matter continuum, God must cause, maintain, organize and manage everything all the time. In this process, he dedicates shares of himself to constituting the individual part-bodies of the world and keeps only the uttermost fringes for himself. What is more, since the world thus created is perishable, not eternal like the *kosmoi* envisaged by Plato and Aristotle,[4] God has to repeat the production of the same optimally designed differentiated world again and again in an eternal sequence of world cycles. He commits himself in an infinite series of new beginnings to make yet another world like the previous one, all this while also planning and agreeing to the inevitable destruction of his new creation. Why would a perfectly rational being in lack of nothing want to do that?[5]

It does not suffice to say that God, like the human sage, is essentially active (Diogenes Laertius 7.124 = *SVF* 3.628)[6] and that therefore it would not be enough for him just to contemplate and imagine a world. We still would have to ask why his thoughts and intentions and eventually his activities take this direction: that there shall be a world in need of constant maintenance and full of imperfect part-bodies, not least of all also vicious people, individuals deliberately acting against God's will and values. Why did God not stay on his own or, if he wanted to create something, make a completely fiery cosmos like the one that exists during the recurrent periods of world conflagration when the God-Matter compound is a single continuous perfect mind.

The answer, I claim, is that God, just like his human creatures, is essentially sociable. Since Matter, the other of the two principles that constitute the world, is by itself unqualified, all qualities of individual God-Matter compounds, i.e. the cosmos as a whole and the part-bodies in the differentiated world, express the qualities of God himself more or less, depending on the proportion to which God is present in them. This tenet is at the heart of the Stoics' adaptation of Aristotle's *scala naturae*, the classical version of the Great Chain of Being. They rank all beings in the cosmos, from stone to star, according to the range of functions the share of God in them, their *pneuma*, is able to perform (Wildberger 2006: 209f., 2008). Those beings that contain more God – humans, and in particular sages, the perfectly developed specimens of humankind and also all individual divinities, such as the celestial bodies – are not only rational but also sociable (*koinōnikoi*), as the Stoic doxography in Diogenes Laertius tells us:

Moreover, the sage will not, they say, live in solitude, for he is naturally sociable (*koinōnikos ... phusei*) and an active agent. (Diog. Laert. 124. 7. = *SVF* 3.628, trans. Gerson and Inwood 1997, slightly adapted)

That the same sociability characterizes humans and God is the basis for the analogy that the Stoic Seneca draws between a human sage temporarily isolated from others and God (here called 'Jupiter') at the time of world conflagration, when all the part-bodies, including the individual gods, have dissolved into the one perfect mind consisting of finest fire. At that stage, God in his function as creative Nature is at rest before starting the next cycle of a differentiated world:

[An interlocutor asks:] 'But what sort of life will the wise man have if he is left without friends when in captivity, or stranded in some foreign country, or delayed on some long voyage, or cast away on a desert island?' [Seneca's reply:] The kind Jupiter has at that time when the world is dissolved and all gods are mingled into one, when Nature ceases her operations for a while and he devotes himself to his own thoughts, and rests in himself. What the sage does is something like that: he retreats into himself and is his own company. (Seneca, *Epistles* 9.16, trans. Graver and Long 2015, slightly adapted to express the theist notion of Nature)

What the analogy is supposed to explain is the self-sufficiency of a sage. Such an individual does not lack anything but still can make good use of many things (Sen. *Ep.* 9.13–14),[7] in particular of other people with whom he wants to be friends. He cares for others not out of self-interest, as a means for achieving some further end, but out of natural inclination; this inclination is so strong that the sage would not even want to live, i.e. see no point in staying alive, if he were to spend the rest of his time without other people (Sen. *Ep.* 9.17). In the same way, God too is a self-sufficient perfect being lacking nothing but still has good use for others because of his sociability. He makes other beings because to him this seems the best use of all his powers and rationality.[8]

Most often, God's sociability is described in terms of his love for humans and gods, with whom he wants to be friends. However, a case can be made – and has been made – against the assumption that Stoicism must be understood as anthropocentric (e.g. see Velayos Castelo 1996 and Holland 1997). I do not wish to repeat the work of others. Instead, I will develop a non-anthropocentric reading of the kind of sociability that characterizes both Stoic sage and God.

Stoics called God also Providence and Nature to refer to aspects of his activity as a sociable being. As Nature, God is that which makes the cosmos a living thing and constitutes individual living part-bodies: Nature creates them with a specific biological make-up and natural functions, assures their nutrition and growth and ultimately their proliferation through further procreation by such living things themselves. As Providence, God is the benefactor of the cosmos as a whole and humans in particular, taking care of the beings he has created as Nature. The connection between these aspects is apparent in a passage from the account of Stoic theology presented by the character Balbus in Cicero's dialogue *On the Nature of the Gods*.

> Zeno, then, defines nature thus: he says that it is a craftsmanlike fire which proceeds methodically to the task of creation. For he thinks that creating and producing are most characteristic of a craft and that nature (i.e., the craftsmanlike fire, as I said, which is the instructor of all the other crafts) accomplishes the same sort of thing as our hands do when they are used in human crafts, but much more skillfully. And on this theory, nature as a whole is craftsmanlike, because it has a kind of method and path to follow; [58] but the nature of the cosmos itself, which constrains and contains all things in its embrace, is said by the same Zeno not only to be craftsmanlike but, to put it directly, a craftsman, since it looks out for and is provident about the utility and convenience of all (*consultrix et prouida utilitatum oportunitatumque omnium*). (*Nat. D.* 2.57f. trans. Gerson and Inwood 1997: 150, slightly adapted)

Craftsman and Nature ('the nature of the cosmos') appear as creative forces: 'creating and producing are most characteristic of a craft', while Providence 'looks out for and is provident about all kinds of usefulness and convenience' (reading *omnium* as an adjective attribute with the translators Gerson and Inwood) or rather, as I would suggest, 'the utility and convenience of all' (reading *omnium* as a noun and genitive attribute). What Balbus describes is a God creating all kinds of things and making provisions to keep them all in their respective optimal condition. Since Providence in particular is also identified with God's volition or *boulēsis*, his active, motivated, and deliberate activity directed at what appears good to him (Chrysippus in Calcidius, *Commentary on Plato's Timaeus* 144 = *SVF* 2.933),[9] we can conclude that this creative activity and caring for all things is deliberate, intentional, and thus something which God himself values highly.

If God wanted only the company of other rational beings with whom he could enter into a friendship in the proper Stoic sense, which is only obtained between sages,[10] he could have chosen to make a differentiated world more like the outer space of the Stoic cosmos, the fiery ether populated by perfectly rational divinities (e.g. see Banateanu 2001: 155–97). However, he chose otherwise and did not make only individuals at the top of the *scala naturae*, to hang out with his buddies, so to speak. Rather he exerts himself to squeeze out as much life and existence from every notch and corner of the cosmos, even out of the murkiest, innermost part full of the Matter-rich passive elements earth and water, in order to make as many part-bodies as possible. Stoics were criticized by their contemporaries for arguing that God is present even in such lowly things as dirt or worms (Wildberger 2006, vol. 1: 16 and vol. 2: 483 n. 105). Yet this one can turn around to say that the divine Craftsman cares so much for even these creations, that he does not hesitate to maintain them in existence by permeating them and integrating them into the differentiated world, just as he permeates and integrates humans and gods.

Stoics famously called the differentiated world one big *polis*, a cosmic state and dwelling for humans and gods (e.g. Cic. *Nat. D.* 2.154).[11] Accordingly, one might argue that all other part-bodies of the cosmos were created for the convenience of its divine inhabitants. However, the heavenly bodies give more than they require for themselves. To maintain their life and fuel their fire they feed on exhalations from the lower spheres, notably the oceans and the Earth (*SVF* 2.658–63).[12] Even this need for sustenance would be greatly diminished if they would not be busy moving in regular circles, emitting their fire to sustain plants and animals and keeping the cosmos in balance. To sustain the gods, then, it was not necessary to make humans, animals, plants, let alone material shaped in form of manifold minerals, waters or geological formations. Even humans could have been kept alive and healthy in some less elaborate way, and one could argue that if God would have had more of himself left and not used shares of himself on other life forms, he might have been able to make humans perfectly rational right away, just as he did it with the stars, to whom he imparted a larger share of himself.

God is described as aiming not only at a multitude of beings, but also at variety: he makes individuals, as many as he can, and each single one different from the other, as Seneca notes with admiration, while referring to God in his function of Craftsman and Nature:

> Look around at the bodies of everything. Not one of them is without its characteristic color, shape, and size. There are many reasons for marveling at the divine Craftman's creativity (*diuini artificis ingenium*), and they include, in my opinion, the fact that in the vast supply of things nothing ever turns out to be just the same. Even things that seem alike are different when you compare them. He has made numerous kinds of leaves, each one with its own distinctive property – numerous animate creatures, yet not one of them is exactly the same size as another, or without at least some difference. (Sen. *Ep.* 113.15–16, trans. Graver and Long 2015, slightly adapted)

There would not have been any need to differentiate all these creations if they were made just in order to provide material comfort to humans and god. If the only reason for the existence of pigs, sheep or vines is to feed and clothe humans, why make each of them a uniquely designed individual?

Nor does it seem that God made the differentiated world just for his own and his rational companions' entertainment; for them to have a variety of toys, as it were, to play with and enjoy. In that case, he could have limited himself to giving gods and humans images and ideas of such things, something like eternal entertaining dreams. Providing a virtual reality would have been sufficient, without causing such things to actually exist. Even existent things could have been designed differently if their sole purpose were to entertain higher life forms. God could have made such lower life forms and other objects like puppets or pawns for himself or his rational companions to move as they like. Instead, he gave non-rational beings and even lifeless things as much independence as was possible and turned them into what Paul Taylor calls teleological centres of life. They all have their own intentions and goals or at least natural dispositions to move in particular species-specific ways (Taylor 1981; Wildberger 2006, vol. 1: 342).

Rather, gods and humans benefit from the existence of this world with all the many beings in it because it allows them to be perfectly good in the Stoic sense, i.e. always to be beneficent. Like God himself, they make good use of the world to practice their inborn beneficence and sociability. Like him, they want there to be more individuals than just themselves and that these individuals thrive as best as it is possible in a world created out of inert Matter. Mortal rational animals must fulfil some basic material needs in order to remain in existence, but that is not what they live for. They are not altruists in the sense that they would put the wellbeing of others before their own, it is rather that their own wellbeing consists in serving others because in this they are following a strong natural urge, the one urge that aligns them most perfectly with their divine creator and leads them to what is best for them. They flourish by being beneficial, by being instrumental to the wellbeing of others. Unless they live for others, they do not live for themselves (Sen. *Ep.* 48.2), or as Epictetus puts it:

> As a universal rule, [God] has designed such a nature for the rational animal that it cannot attain any of its proper goods if it does not bring forth something beneficial for the community (*koinon*). It is in this way that doing everything for one's own sake is no longer unsociable (*akoinonēton*). (Arrian, *Discourses of Epictetus* 1.19.13–14, my translation)[13]

Sociability in the Stoic sense, then, is an intrinsic and permanent desire characteristic of God and other rational animals, which aims at fostering the existence of as many other things and beings as possible, letting them flourish, each in its own individual way, and keeping them in existence as long and in as good a condition as possible.

Beauty

Full sociability obtains only in completely rational animals, those that have the greatest share of God and, in the case of humans, have developed it to perfection. However, in attenuated form this essential characteristic permeates the whole cosmos and occurs in diminishing degrees in all beings, which all contain at least a small share of God. Since there is nothing outside the cosmos from which the cosmos could draw supplies, the cosmos is both the product and the resource for creation. This means that also the single creatures, the part-bodies within the cosmos, serve as resources for each other and, at least in this sense, are useful for other created things. Continuing the account quoted above (p. 66) in which the Stoic Balbus describes God's activities as creator and provider for all kinds of individuals, Cicero now looks at Providence from the viewpoint of the whole cosmos.

[T]he principal concern of this Providence and her greatest preoccupation is, first, that the cosmos be as well suited as possible for remaining in existence, second, that it lack nothing, but most of all that it should possess surpassing beauty (*pulchritudo*) and every adornment (*ornatus*). (*Nat. D.* 2.58, trans. Gerson and Inwood 1997, 150, slightly adapted)

Like the individual part-bodies, the cosmos too must be maintained in existence and well provided for, and then also have beauty. Since there is nothing outside the cosmos, there can be no question of the cosmos as a whole receiving provisions from somewhere else. Providing for the cosmos so that it lacks nothing can only mean that it is managed internally in such a way that it is sustainable. Similarly, since there is no outside observer, the beauty of the cosmos can only be desirable for God, the cosmos itself, and the beings in it. The purpose of its beauty is intrinsic to the cosmos.

Beauty reoccurs more than once in Balbus' account as evidence for the existence of gods; it is a sign of divine agency of the kind that, as I have argued, is an expression of God's sociability. Everything is managed most beautifully (*pulcherrime*) by nature (*Nat. D.* 2.81), an activity manifest in the divine artwork of the 'wonderfully adorned and most beautiful' (*ornatissimum et pulcherrimum*) heavens (*Nat. D.* 2.93, cf. 2.95) with that innumerable multitude of fiery stars (*Nat. D.* 2.92). But beauty is everywhere in the world for us to observe the workings of a rational mind. One must only open one's eyes and look at 'the beauty of the things that we claim to have been constituted by divine providence' (*Nat. D.* 2.98). In the description of this beauty, Balbus repeatedly emphasizes the variety and multitude of what has been created:

[all kinds of plants] all of whose (*quorum omnium*) unbelievable multitude is set off by insatiable variation ... Now, the sea: How great is its beauty, how impressive the overall view of it, how great the multitude and variety of the islands ...! (*Nat. D.* 2.98. 100, my translation)[14]

Given that beauty occurs in such contexts and is attributed greatest importance, as a primary aim of God in his function as caring Providence, I would suggest that cosmic beauty is one type of the Greek concept *kalon*, the beautiful-and-honourable (in Latin: *honestum*), which is an essential property of anything the Stoics acknowledge as 'good' in the proper sense of that term. Given further that what is good is essentially either benefit or nothing else than benefit, according to the Stoics,[15] I would also suggest that cosmic beauty is the property of the cosmos which ensures optimal benefit of both the cosmos as a whole (in the sense of its optimal diversification in number and kind, its sustainability and its maximal durability) and of the individual part-bodies in it (as that which brings about their flourishing to the utmost degree possible). Finally, given that *kalon* has an altruistic component in general ancient Greek usage, in Stoic ethics, and in its Latin translation *honestum* and that it is opposed to self-serving advantage (*sumpheron*, or *utile* in Latin), I would suggest that cosmic beauty too involves some kind of beneficent sociability or instrumentality for others.

The Stoics famously claimed that only what is *kalon* is good, i.e. that beauty in this sense is a necessary condition for something to be good, while every good thing is also beautiful-and-honourable. Our sources are fragmentary here, but there are good reasons to assume that (a) the anonymous definition of beauty rejected in Plotinus' treatise *On Beauty*, which corresponds to similar definitions of specific kinds of beauty, is the basic formula used in various extant Stoic definitions of specific types of beauty (*kallos*) and that (b) the Stoic 'beautiful-and-honourable' (*kalon*) should be understood in these terms. According to the definition cited by Plotinus, beauty is 'the symmetry of the parts towards each other and towards the whole' (λέγεται μὲν δὴ παρὰ πάντων, ὡς εἰπεῖν, ὡς συμμετρία τῶν μερῶν πρὸς ἄλληλα καὶ πρὸς τὸ ὅλον) (Plotinus, *Enneads* 1.6.1 in Henry and Schwyzer 1964: 92, lines 21–22).

That this is a Stoic definition is shown by Aiste Celkyte in a paper on 'The Stoic definition of beauty as *summetria*' (2017) with ample evidence and a comprehensive overview of previous scholarship. Celkyte also argues convincingly that this definition expresses a functional conception of beauty, i.e. that the criterion for assessing the beauty of a thing was the degree to which its structure, the symmetry between the parts among each other and with regard to the whole, is contributive to the respective purpose or *telos* of the object. Celkyte's claim can be applied to the cosmos as a whole: cosmic beauty too is a property that contributes to the cosmos's achieving its specific purpose. And as we have seen above, the purpose of the cosmos coincides with the aims of Providence, the expression of God's sociability. Just as goodness (*aretē*) in a rational individual consists in the maximal contextualization and integration of that individual's natural strivings, values and sub- or part-aims (*hupotelides*),[16] so the goodness of the cosmos consists in achieving the goal set by Providence, and beauty is the property of the cosmos that contributes to it.

Providence aims at creating as many things as possible and promoting the interests of each, in such a way that the whole, namely the cosmos, contains all the provisions needed both for the maintenance of the individuals and for its own continued existence. Now, if there is a multitude of diverse individuals, each with their own interests, conflicts arise. The woodpecker wants to live in the tree, the beaver fells it. Animals kill plants and fellow animals to eat them, and humans employ beasts for all kinds of purposes. What I have taken, is no longer available for you. Beauty, I wish to suggest, is that feature of the cosmos, which reduces such conflict and turns it into harmony or even, to use a modern term, synergy. Cleanthes claims that it is God's achievement to integrate human badness and fit disparate forces, positive and negative, into a harmonious whole:

> For thus you have fitted together all good things with the bad,
> so that there is one eternal rational principle (*logos*) from them all.
> (*Hymn to Zeus* in Stobaeus, *Anthology* 2.1.12; Wachsmuth 1884: 26, *SVF* 1.537; trans. Gerson and Inwood 1997: 140)

'Mutual harmony among all parts of the universe' characterizes 'the tremendous sympathy, agreement and interconnected relationships' between the individuals and parts of the cosmos that, according to Balbus in Cicero's *On the Nature of the Gods*, is constitutive to its function and necessary for the existence of both the parts and the cosmos as a whole (*Nat. D.* 2.19, trans. Gerson and Inwood 1997: 143). The part of the cosmos that is most stable and tranquil, and contains the individuals that live longest in the differentiated world is the supra-lunar ether. This same sphere is also the most beautiful, characterized by harmonious interconnectedness between all parts and the whole: such are the 'stable movements of the heavens and well defined hierarchical structures among the stars' and to such a degree 'are all things (*omnia*) interconnected and conjoined with each other' up there (Cic. *Nat. D.* 2.19, my translation). It is the beneficent activity of Providence that not only structured them in this way but also effected in this manner that there is no conflict between them. As Seneca insists:

> [T]hings propelled by chance frequently fall into disorder and soon bump into one another, whereas this rapid motion continues without collision, obeying an eternal law and carrying along with it such a quantity of things on land and sea, so many lights shining and reflecting in an ordered arrangement. (Seneca, *On Providence* 1.2, trans. Ker 2014)

We could thus say that cosmic beauty is the ideal implementation of a universal utilitarianism encompassing all things there are and also the whole. The divine Craftsman and Manager (*dioiketēs*) squeezes out a maximum of utility and convenience for all (Cic. *Nat. D.* 2.58).[17]

However, it is a utilitarianism of a very peculiar nature, for three reasons: first, it is individualist, but also includes the whole, the cosmos, as yet another individual. Beauty is not only the symmetry between the individual parts, but also between all the parts and the whole, and in this sense, holistic as well.

Second, the incompatibility usually perceived between individualistic and holistic approaches – either self-interest of the individual or loss of individual values within the whole – does not obtain. On the one hand, the cosmos is an individual just like all the part-bodies; on the other hand, the part-bodies benefit from the well-being of the whole. There can be no beauty only with parts in harmony among each other since there also needs to be a purpose by which that harmony is measured, and that purpose is provided by the whole. If the whole does well and is beautiful, then the parts are in a good and beautiful state too, not only because they are parts of a cosmic organism, but also because parts are not something else than the whole (e.g., Sen. *Ep.* 95.52).[18] Nevertheless, the parts and their respective interests do not resolve into the whole. They are also existent things by the same criteria as the cosmos and, again, by the same criterion of having a constitutive share of God also individual teleological centres of activity. What is more, the interests of the part are constitutive of the whole and its beauty. Plotinus acutely remarks and then criticizes that according to the Stoic definition only a compound, something composed of parts, can be beautiful. The existence of individual part-bodies in the cosmos is a necessary condition for the cosmos to be beautiful. This is so because creating individuals is the raison d'être, the constitutive purpose and natural function of the cosmos as a whole. There would be no point in having a cosmos if it did not have such parts with individual interests, and without them, the cosmos would not be good. One might even go so far as to say that if he did not create such a cosmos, God himself could not be beautiful or perfectly good. He would then have powers and dispositions, notably his rationality and his sociability, but not make use of them and miss the aim that is set for him with these properties of his.[19]

Third, and most importantly, Stoic beauty-utilitarianism is peculiar with regard to the kinds of interests it negotiates. It is not just the best possible compromise between clashing interests of already existing things. Rather, it maintains that the interests and preferences of each were designed together with the cosmos as a whole in such a way that the largest possible number and variety of individuals could all flourish in a mutually supportive, optimal way. Here, again, we have a holistic trait, in that the individual's biological make-up includes its integration into a whole and the alignment of its interests with the interests of the other and the whole. However, this is not a Stoic variation of *Brave New World*, in that preferences of individuals would be suppressed together with their potential for flourishing already in the design of the beings when they are created. Rather, their property of being holistically aligned is an essential element of their flourishing. After all, it is none other than the expression of God's sociability in his human and non-human creatures. Just as God's good consists in being beneficent and instrumental to other beings, so the individual part-bodies' good includes being useful for other part bodies. Plants, animals and other natural entities are instrumental for each other because such is their intrinsic desire as children and shares of God, and they achieve more of that good because they have been designed to fit into the beautiful whole and be in agreement with the other parts of it. Is is a curse not to be instrumental to others, to be disconnected from them. Epictetus writes that if crops had perception, they would be sad when no one would harvest them. 'Know that in this manner it is also a curse for a human being not to die: that's similar to ripening but not being harvested' (Arr. *Epict. diss.* 2.6.12–13, my translation). We have already seen that instrumentality among the part-bodies is not just bottom-up but also top-down: like God, the mortal gods care for lower creatures, also plants and animals, and the perfect human sages care about human fools as well. Seneca, too, alludes to the heavens' life-giving activity in the passage from *On Providence* quoted above (p. 71) the stars enjoy unhindered movement, but the purpose of that movement is the maintenance of life and flourishing on Earth. According to the principle that entities with a higher proportion of God in them also have a higher proportion of God's properties, instrumentality might increase in the higher rungs of the *scala naturae*. Fully developed humans and all gods are by nature even more sociable and more willingly instrumental than other creatures; because they possess reason, they are also more capable to understand the social nature of the whole cosmos, to contextualize their own existence in order to take account of the interests of others, and to both realize and accept their inborn desire to benefit others.

A new environmental ethic?

By overcoming the dilemma of holism vs individualism that bedevils discussions in environmental ethics and by defining individual interest and even more so human preference as a desire to be instrumental, Stoic cosmic beauty-utilitarianism may thus suggest starting points of a new environmental ethic. One answer to the dilemma already offered has been the application of virtue ethics to environmental issues.[20] The Stoic theory outlined above could ground such an approach by providing reasons why environmental virtues and the concomitant behavior may be something individuals should strive for. The theory also suggests criteria for redefining interests and values for distribution and defines conflict in a new way. Utilitarianism traditionally assumes necessary clashes between individual interests. In stark contrast to this, cosmic beauty functions because it is built from altruism, so to speak; all part-bodies are not only made *because* of sociability but literally *out* of sociability. The Stoic concept of beauty takes account of the interconnectedness of individuals without disregarding their needs, and especially the needs of the weakest. According to the Stoic model, utilitarian considerations should provide what enables as many beings as possible to flourish as fully as possible, for example as set out for humans in the Capability Approach of Amartya Sen and Marta C. Nussbaum (Nussbaum 2011).[21] As concerns us humans, the Stoic model would, however, also provide a clear teleology for such an approach: in developing her or his various capabilities, human beings learn to become increasingly instrumental in a way that does not stop at the boundary between human and non-human – or sentient and non-sentient, for that matter. Fully flourishing sages would also care for the different non-human 'societies' of which they are part as citizens of the cosmic *polis*, such as the universe, the biosphere of planet Earth or the local ecosystems they live in. The strength, power and means humans acquire by developing their individual capabilities will have a meaning, and goodness, only in the service of all the other beings and the community as a whole. If a human being fails to be instrumental, unnecessarily exploiting and harming other beings and destroying diversity rather than promoting it, then this is a sign of weakness and unhappiness, that the person or society in question has not been able to develop their sociable potential and not yet achieved the goodness and internal beauty of which it would be capable.

Space precludes developing these ideas more fully here. I will conclude with a few considerations about the viability of taking my interpretation of Stoic cosmic beauty-utilitarianism as a starting point for a theory that could inform ecological action in the twenty-first century, given that the physical and metaphysical underpinnings of ancient Stoicism are no longer tenable. For a theist, the identification of natural desires deriving from one's biological make-up with our personal values and goals need not be a natural fallacy: God orders persons to be like that by designing them in this way, and it is good for them to follow God's commands. The same can be said about attributing values and goals to the cosmos as a whole, which becomes problematic for an atheist, for whom the only origin of value can be the minds of valuers, beings with intentions and desires, like animals or humans. There are, however, other ways in which we could imagine the cosmos to be sociable in a non-intending way, such that our own sociability is the result of natural principles characterizing and constituting our environment. For example, we might look at the world as a complex structure with manifold and interrelated natural functions, in the naturalist sense of the word. Even though there is no eternal stability, as there is none in the Stoic cosmos either, at least the world seems to be a structure capable to sustain, at least for a time and a certain degree, perishable diversity and a multitude of beings. It also seems to be the case that this structure not only fosters diversity, but also arises from the 'sociability' of the diverse individuals constituted through this structure, from particle to galaxy cluster, which interact to form even more complexity and diversity. The same is true of the interdependent individuals in an ecosystems. Evolution drives speciation and thus diversity, and otherness in form of mutation as well as intra-species diversity drive evolution. Most life forms, even cells, thrive in functional and social organizations, and even if we look at the basic emotions not only of humans but all the other sentient animals, it becomes evident that sociability and orientation towards the other is woven into the fabric of even the deepest layers of consciousness (Panksepp and Biven 2012). We can conclude that there is sufficient material even for a scientifically minded atheist of our times, to begin thinking about the world in terms of a naturalist functionalist teleology of sociability.

This does not solve the problem of is and ought, though. Even if the cosmos were teleologically ordered and beautiful in the Stoic sense, why should a person embrace those cosmic values? Here, we could turn to the Stoic theory of 'attachment' (*oikeiōsis*) and human development, which involves a full process of self-perception and decision making on the basis of a self-evaluation. According to the Stoics, we are not automata driven by a biological programme of instincts, but act because we have chosen to follow the suggestions of our nature and seen that it is rational to do so. Accordingly, it should be possible to adapt the Stoic theory if we share the assumption that it is, in fact, part of our biological make-up that we have ethical instincts or intuitions (Callicott 1989) and that a fundamental and very strong such intution is sociability of the kind described above. What the epistemic status of such an assumption would be depends on the world view of the modern theorist. For the current purpose, one answer a Stoic would give must suffice: 'Ask yourself! – But whatever answer you give, always consider whether you say this out of weakness and inconsistency, whether this answer is in agreement with your other ideas about yourself and what is good for you, and whether you really are as vulnerable, helpless and dependent on resources as it may seem to you.'

CHAPTER 6
SOME ANCIENT PHILOSOPHICAL AND RELIGIOUS ROOTS OF MODERN ENVIRONMENTALISM
Robin Attfield

Introduction

Besides its recent origins and cultural influences, modern environmentalism draws on a wide range of biblical and classical ideas and themes, such that one can reasonably affirm that, in part, its roots lie in the texts of the ancient world. In this chapter, some of the more significant of these sources are discussed, together with comments on the interpretations placed on them and on the uses to which they have been put in the modern era. I will argue that several of these ancient sources have contributed to important strands in contemporary environmentalism.

There is a surprising amount of material to consider; hence many texts that might be regarded as influencing modern environmentalism must of necessity be omitted. For example, tragedians such as Aeschylus and Sophocles often wrote of the Earth as mother (Aeschylus *Libation Bearers* 45, 127; Sophocles *Oedipus at Colonus* 506–7), and also of the perils of human hubris (Sophocles *Oedipus the King* 19, 163–5, 245 and *passim*), as did the historian Herodotus (*Histories* 3.29, 7.35, 7.238 and 8.5). But it would take more than a chapter (and perhaps more than a monograph) to fully survey environmentally relevant themes in the ancient tragedians and historians alone. Nor is there space to write of ancient writers of epic, didactic or lyric poetry, of comedy or of satire, despite the potential relevance of some of their themes. The treatise *On Airs, Waters, Places*, attributed to Hippocrates, can be seen as pre-figuring the late twentieth-century subject of bio-climatology, and was recently hailed by Anthony Capon as an ancient anticipation of modern ecological studies, but the corpus of ancient medical writings must here be set aside, together with bio-climatology, for others to study.[1]

In this chapter I will focus instead on the writings and influence of ancient philosophers, and of key ancient religious and theological writings. Much more than literary works, these have frequently been the foci of attention and influence, and sometimes the targets for criticism, of modern environmental thinkers. I will begin with a brief survey of ancient philosophers and religious writers, who have been significant voices on this topic.

Ecology and Theology in the Ancient World

Some key ancient thinkers

In this section, I present a broad range of philosophical and religious ideas and texts that might be seen as potential influences on modern environmentalism. These include, *inter alia*, expressions of belief in cosmic purposes and cosmic cycles, in the light of modern theories such as the Gaia theory, where the claim that the Earth is a self-maintaining system with multiple systems adapted to the maintenance of the whole (Lovelock 1979), has sometimes been interpreted as appealing to an inbuilt natural teleology.

Philosophical texts

Among the Presocratics, Anaximenes' notion of cosmic cycles, with processes such as the condensation and rarefaction of air as their base, could have served as a model for cycles such as the nitrogen and carbon cycles central to ecological science, but has not received sufficient attention from subsequent thinkers to exercise such influence (DK 90–95; see also Kirk et al. 1983). However, Empedocles' theories of the evolution of living creatures from earlier forms did become a conscious model for later ancient evolutionists such as Lucretius (see On the Nature of the Universe, book 5), even though the latter did not agree with Empedocles' four-elements theory (DK B21 & B23). This was also subsequently adopted by Charles Darwin, whose related theory of evolution by natural selection has indirectly influenced environmentalism, not least through conveying both the co-dependence and the vulnerability of Earth's species.[2]

In Plato's dialogues, Socrates in the *Phaedo* expresses disillusion over Anaxagoras' failure to carry through his fundamental belief that everything is ordered by *nous* (Mind), and seeks to foster the study of purposiveness, whether external or inbuilt (*Phaedo* 97b7-c1). Final causes continue to figure in debates about the functions of interacting species within modern ecological science (see Belshaw 2001: 148–9; Shrader-Frechette 2001: 307–10).[3] The claim in Plato's *Phaedrus* that 'Soul, considered collectively, has the care of all that which is soulless'(*Phaedrus* 246b)[4] is cited by Passmore as one of the foundation texts to support contemporary belief in human stewardship of nature (Passmore 1974: 28). But the dialogue that has proved most influential has been *Timaeus*, with its claim that the world is a living creature (*Timaeus* 30c), and Plato's account of its ordering by the Demiurge, the cosmic architect (*Ti.* 29a). These perspectives can be understood as contributory factors in the development of holistic environmental theories such as the Gaia theory of James Lovelock (Lovelock 1979), of the holistic ethic of Aldo Leopold (Leopold 1949), of the views of Deep Ecologists (Naess 1973) and of eco-holists such as J. Baird Callicott, who also sought to interpret Plato's ethical holism (Callicott 1989).

Callicott draws attention to the parallel between Aldo Leopold's comparison of land to an organic body or human society and Plato's comparison between a human soul, body and society in *Republic* 4.444a-e, both, according to Callicott, having similar 'structures and corresponding virtues' (Callicott 1989: 28; see also Callicott 1999:

333–5). Thus, '[t]he goodness of each is a function of the structure or organisation and the relative value of the parts or constituents of each is calculated according to the contribution made to the integrity, stability and beauty of the whole' (Callicott 1989: 35). This phraseology echoes the triad employed by Leopold in his 'land ethic' to appraise the rightness of actions (Leopold 1989: 224–5). Callicott further draws attention to Plato's account in *Gorgias* 503d-507a of the fitting relation between a thing's goodness in terms of its parts and the whole that those parts comprise (Callicott 1989: 28 n.35). He goes on to remark that Plato seems to regard individual human pain and suffering with complete indifference, making the interest of the community the sole criterion of value, and to regard individual organisms in the same light (Callicott 1989: 28–9). Callicott proceeds in this same essay to endorse Plato's view of the value of pain and the value of individuals as good or bad instrumentally rather than intrinsically; but in a later article, now printed in the same collection, he retracted this stance, preferring the animal welfare and 'biosocial' stance of Mary Midgley, and adjusting his account of Leopold accordingly (Callicott 1989: 49–59).

Aristotle revived Socrates' theme of purposefulness, expressed in Plato's *Phaedo*, by including final causes, alongside material, formal and efficient causes, as the four kinds of cause in his metaphysic (*Physics* 194b-195b; *Metaphysics* 981a-983a). Aristotle's rejection of Plato's theory of forms (*Metaph.* 991a8, 1033b26-28; *Ph.* 259a8) and his particular kind of empiricism can be seen as encouraging the empirical aspects of modern scientific methodology, while his detailed studies of animals, continued by his disciple Theophrastus, have not only served as an inspiration for modern biology, but also as attesting to the wonder and the value of each and every creature (*The Parts of Animals* 645a26-27; see Nussbaum 2006: 348 and Johnson 2005: 290). An apparently anthropocentric passage in Aristotle's *Politics* (1256b 15–22) suggesting that everything in the world exists for the sake of humanity has led to a recent debate about whether or not Aristotle was an anthropocentrist (Sedley 1991; Wardy 1993). However, there are grounds to doubt this ascription as we shall see when we discuss anthropocentrism in more detail below. While doubts are also in place about whether Aristotle could have endorsed the attitudes of modern ecologists, because he regarded nature as permanent and ultimately invulnerable (*Meteorology* 352b16-20), there are good reasons to suppose that he would have welcomed modern ecological science (*Part. an.* 645a26-27).

The Stoics, with their belief in everything existing for a purpose (Castelo 1996), have also sometimes been accused of anthropocentrism, possibly with greater justification, for example by Passmore, who cites Chrysippus and Cicero's character Balbus in *On the Nature of the Gods* to support this view (Passmore 1974: 14–15 and 18–19). But more recent studies have cast doubt on this interpretation (see Castelo 1996; Holland 1997 and, in this volume, chapters by Jedan and Wildberger). Also, as we shall see below in connection with the tradition of 'co-operation with nature', the first-century BCE Stoic Posidonius seems to have adhered to the view that the role of humanity is to enhance or perfect the cosmos, a view with important modern implications. This certainly is the view of Passmore who writes,

> This attitude to nature [i.e. cooperation] is sometimes formalised as a metaphysics. As such, it may originate with the Stoic Posidonius, writing in the first century BC, if this is how we are to interpret his teaching that man's task is 'to live contemplating the truth and order of all things and *doing one's part in helping* to establish that order.' (Passmore 1974: 33, his italics, trans. Nock 1959: 12)

This observation rather reduces the credibility of Passmore's charge against the Stoics of arrogance towards nature, since he does not appear to notice the tension between his interpretation of the Stoics at pages 14–17 and his reference to Posidonius on page 33.

As already alluded to, the Epicureans' belief in both cosmic evolution and in the evolution of living creatures (Lucretius, On the Nature of the Universe, book 5), and their rejection of teleology in nature (*DRN* 5.124–234), have served to foster Darwinian studies of evolution, and thus, indirectly, the science of ecology. Of equal importance has been their presentation of humanity as vulnerable to natural forces, with no supernatural guarantees of immunity (*DRN* 5.196–226).

Perhaps slightly more controversially, the question of vegetarianism was addressed by the Neoplatonists (third to fifth centuries CE). Their awareness of the links between human beings and other animals led figures such as Porphyry, in *Concerning Abstinence* (IV, 20), to pioneer vegetarianism (Passmore 1974: 115, 178). While not all environmentalists are vegetarian, some environmentalist policy proposals advocate vegetarianism as one contribution towards reducing human impact on the earth's resources (Nolt 2015), and there is an increasing move towards reducing meat consumption in some sections of Western society.

Religious texts

In the Hellenistic Gnostic tradition, the Hermetic tract *Asclepius* (second century CE) asserts that 'God willed that the Universe should not be complete until man had done his part' (*Asclepius* 1,8; see Glacken 1967: 146; Passmore 1974: 33). This text (like the *Phaedrus* passage mentioned above) may well have fostered belief in the human role in enhancing or perfecting the planet, a belief highlighted more recently by Passmore (1974: 32–9). Passmore contrasts the theme of humanity helping with the task of creation, which he finds probably in Posidonius and certainly in *Asclepius*, with the stances about creation of both Plato's *Timaeus* and the Old Testament, in which, in his view, there is no human role in creation. This theme, he controversially claims, does not resurface until the works of German metaphysicians such as Johann Gottlieb Fichte in the late eighteenth/early nineteenth century (Passmore 1974: 33–4).

The relevance of biblical texts to contemporary environmental concerns has been much debated in environmental literature over the past few decades (for example Berry 2006, Preston 2009 and Rolston 1999). Significant Old Testament passages that offer a positive view of the physical creation include Psalm 104, with its belief that wild places were created for beasts such as lions, and the sea for sea creatures, and texts about the welfare of animals such as Prov. 12:10, which asserts that 'A righteous man regardeth the life of his beast.' (KJV)

The writers of the New Testament are less explicit, but largely take Old Testament stances for granted. Jesus recognizes the independent value of 'the lilies of the field' and 'the birds of the air', adding that human beings are of much greater value (Mt. 6.26–29). Paul regards the whole of creation as involved in God's plan of salvation (Rom. 8.21–23). As we shall see, modern environmental theorists differ in the ways that they interpret both the Old and New Testaments. Many scholars, however, conclude that the Bible supports the view that human beings are God's stewards or trustees, charged with the care of the physical earth. Such interpretations have influenced the outlook of many modern Jews and Christians, and also Muslims, who interpret the Qur'an along similar lines (Haq 2001).

Some of the Church Fathers have been cited as sources of inspiration by modern environmentalists. Paul Santmire, for example, finds a neglected but ecologically promising motif in the second-century CE Christian theologian Irenaeus, who wrote of the whole cosmos being renewed in the end times: 'the whole creation shall, according to God's will, obtain a vast increase, that it may bring forth fruits such ... [as] Isaiah declares [30: 25f]' (Irenaeus, *Against Heresies* 5.34.2; see Santmire 1985: 35–44). Likewise, C.W. Hume, founder of the Universities Federation for Animal Welfare, cites with approbation a teaching about animals purporting to come from Chrysostom – 'Surely we ought to show them great kindness and gentleness for many reasons, but, above all, because they are of the same origin as ourselves' – in a discussion of the frequency with which representations of saints are associated with animals (Hume 1957: 26). Andrew Linzey refers to the same teaching in a passage about Christians of the patristic age who recognized the claims of animals and championed them (Linzey 1987: 17). Tradition holds that Basil the Great (fourth century CE) prayed for 'the humble beasts who bear with us the heat and burden of the day' suggesting that the attitudes of the book of Proverbs may have persisted into the patristic period (attributed to the Liturgy of St Basil; see Moss 1961: 5; Passmore 1975: 198).[5] However others, such as Origen (third century CE), seem to have adhered to a more anthropocentric stance, writing of 'the wildest of animals' that 'even these animals were made for the exercise of the rational being'. (Origen, *Against Celsus* IV: 75, trans. Chadwick 1953; see Passmore 1974: 16, Glacken 1967: 185–6). Nevertheless, the belief that humanity has the role of adorning and perfecting God's creation seems to have been held widely among the Church Fathers, including Ambrose (fourth century CE) and Theodoret (fifth century CE) (Glacken 1967: 192, 299 and 300), while implicit belief in the human role as stewards of nature was also found among them (Glacken 1967: 300–1).

The remainder of this chapter will investigate a number of themes central to modern environmentalist approaches, and to issues around which criticisms of one or another group of these ancient thinkers cluster.

Ecology and Theology in the Ancient World

Anthropocentrism and nature's intrinsic value

There are at least two forms of anthropocentrism, metaphysical or teleological anthropocentrism (which maintains that everything in nature exists for the sake of humanity), and axiological anthropocentrism (which holds that nothing is of value except human interests and whatever promotes or upholds them).[6] It is sometimes maintained that Aristotle, the Stoics and the Bible were committed to teleological anthropocentrism (for Aristotle and the Stoics, see Sedley 1991: 179–96; for the Stoics, see also Passmore 1974; for the Bible, see White 1967: 1203–7). Although this teleological perspective has often resulted in adherence to a narrower axiological anthropocentrism, the latter view is often held independently of any metaphysical or teleological basis. Belief in nature's intrinsic or non-derivative value is incompatible with the second kind, and adoption of this view casts doubt on adherence to the first kind as well.[7]

Prior to Aristotle, these issues were addressed only obliquely in Greek literature. The widespread belief in the presence of deities resident in the earth or the heavens, to be found in the Homeric poems, in the lyric poets and tragedians and in some of the Presocratic philosophers, is difficult (albeit not impossible) to reconcile with adherence to anthropocentrism of either kind, that is with the view that natural entities have nothing but instrumental value, or are made solely for the sake of humanity. However, explicit claims about nature's independent value are hard to find, as opposed to statements about the aesthetic value of earth, heaven or humanity; Sophocles' chorus in Antigone, for example, declare that nothing is more *deinos* (terrible/marvellous/clever) than man (*Ant.* 334, see Attfield 2017: 1–8).

When it comes to Aristotle, there is disagreement over whether he regarded everything as arranged for the sake of human beings, as the following passage from his *Politics* seems to suggest:

> After birth, plants exist for the sake of animals, and the other animals for the sake of humans – domesticated animals for both usefulness and food, and most if not all wild animals for food and other assistance, as a source of clothing and other utilities. If, then, nature makes nothing incomplete or pointless, it is necessary that nature has made them all for the sake of humans. (*Pol.* I 3, 1256b15-22, trans. Carone 2001: 74)

In this passage Aristotle appears to adopt a form of teleological anthropocentrism, but his writing here is in tension with other statements of his, for example, 'If, however, there is anyone who holds that the study of animals is an unworthy pursuit, he ought to go further and hold the same opinion about himself,' which could be taken to suggest that other animals are equally worth of study as human beings (*Part. an.* 645a26-27; see Carone 2001, Nussbaum 2006 and Wardy 1993). It also contrasts with some significant silences, such as the absence of anthropocentric remarks in *Meteorology*. Much turns on how *Metaphysics* Lambda, 10, 1075a16-17 – on which David Sedley, who ascribes a teleological anthropocentric view to Aristotle, centrally relies – is interpreted (Sedley

1991: 179–80, 194–5). While this passage, which treats the universe as analogous to a household with respect to having some form of organization, cannot be discussed in detail here, Robert Wardy has cogently replied to Sedley that no specific form of purposiveness is there ascribed to 'the joint arrangement of everything', let alone the anthropocentric one favoured by Sedley (Wardy 1993: 24). Certainly the fact that humanity is a beneficiary of the terrestrial stages of 'the scale of nature' need not in any case show anything about the purpose of the heavens or of the earth itself.[8] Besides, as Sedley himself admits, Aristotle's silence about anthropocentrism in his *Meteorology* and in his zoological works forms a significant problem for the anthropocentric interpretation (Sedley 1991).

Accordingly, without further discussing the passage from *Metaphysics* Lambda, I am inclined to adopt the view of Wardy (1993: 18–30) and of Carone (2001: 67–80) that Sedley's argument for his anthropocentric interpretation is inconclusive, and that the passage quoted above from the *Politics* may well thus be either an aberration, or possibly not Aristotle's own view at all. Purposiveness in Aristotle would largely consist in the striving of each kind of entity to fulfil its own nature. This is not to deny that for Aristotle the whole of nature embodies a purpose, but it allows for this purpose to consist in each thing complying with its nature and imitating, as far as possible, the highest being.

As the quote above suggests and Nussbaum emphasizes, Aristotle's beliefs about humanity and hierarchy did not prevent him regarding the study of animals as worthwhile (Nussbaum 2006: 348). Monte Ransome Johnson, who himself also rejects a narrowly anthropocentric interpretation, adds that 'Philosophy is … [for Aristotle] a kind of biophilia, which loves and delights in natural objects and their causes … in their own right, not insofar as they can turn a profit' (Johnson 2005: 290). If he is right, then Aristotle may even be regarded as a biocentrist, holding that all living creatures have moral value, and not human beings only.

Teleological anthropocentrism, then, should probably not be ascribed to Aristotle. It has, however, been attributed more widely to the Stoics, for example by Sedley (1991) and Passmore (1974: 14–15), who cites the attitudes of Balbus in Cicero's *On the Nature of the Gods*. His character Balbus affirms that 'Then the earth, teeming with grain and vegetables of various kinds … does she appear to give birth to this produce for the sake of the wild beasts or of men? … It must therefore be admitted that all this abundance was provided for the sake of men … So far is it from being true that the fruits of the earth were provided for the sake of animals as well as men, that the animals themselves … were created for the sake of men' (*Nat. D.* 2.63. 156–8, see Passmore 1974: 196; Shapiro and Curley 1965: 433). However, Carmen Velayos Castelo defends Stoicism against such charges of anthropocentrism. While the Stoics were logocentric (reason-centred), she suggests, they held this stance not about the reason of individuals, but in a holistic manner, since reason is for them the universal nature, pervading both animate and inanimate nature. Hence, even if human beings have greater value than other species, they are only part of what has value, and only the cosmos as a whole is 'perfect' and 'self-supporting' (Castelo 1996: 291–6).

Further, anthropocentric interpretations of Stoicism depend wholly on reconstructions based on works of Cicero such as *On the Nature of the Gods*, because we have no more than fragments of the earlier Greek Stoics. According to Malcolm Schofield, 'The extent of the anthropocentrism of [Cicero's] teleology … varies in Cicero's different formulations. Probably we should take the most careful of them to be the most authoritative … ' (Schofield 1991: 66; see also Castelo 1996: 293–4). The central Stoic claim about the role of man is that man is created in order to contemplate and imitate the universe and our individual natures are parts of the whole. For Chrysippus, the prominent Greek Stoic of the third century BCE,

> And this is why the end [*telos*] may be defined as life in accordance with nature, or, in other words, in accordance with our own human nature as well as that of the universe, a life in which we refrain from every action forbidden by the law common to all things, that is to say, the right reason which pervades all things. (Chrysippus, *On Ends*, cited in *Diogenes Laertius* 7.87–88; see also Castelo 1996: 294)

Indeed, Castelo maintains that 'when Stoics claim that animals were not born for their own sake, but for the sake of Gods and human beings, they may intend that all creatures are adapted to one another's needs' (Castelo 1996: 295). To support this, Castelo draws on Cicero's suggestion that, according to Stoics such as Chrysippus,

> Everything else except the world was created for the sake of some other thing; thus the corn and fruits created by the earth were created for the sake of animals, and animals for the sake of man … man himself however came into existence for the purpose of contemplating and imitating the world. (*Nat. D.* 2.14, 37; see Castelo 1996: 295–6)

The Stoics, then, adopted anthropocentric perspectives only in a qualified sense, in which anthropocentrism was tempered with cosmic holism. This casts doubt on Passmore's negative appraisal of Stoicism and its value for environmental thought (Passmore 1974: 17), and suggests that Stoic thought may warrant further consideration, as has been argued by Alan Holland (Holland 1997: 151–66).

Both teleological and axiological anthropocentrism have been ascribed to the Bible, and to the Judaeo-Christian tradition in general, by Lynn White Jr. 'Especially in its Western form', says White, 'Christianity is the most anthropocentric religion the world has seen … [and has] insisted that it is God's will that man exploit nature for his proper ends' (White 1967: 1205). White's explicit exemption of Orthodox and Coptic Christianity from the full weight of these strictures makes clear that his central target is the Christianity of the West, both medieval and modern. His relative silence about the Bible suggests that he is hesitant about construing it as anthropocentric and as wholly authorizing despotic attitudes to nature, although he does suggest from reading Genesis that 'no item in the physical creation had any purpose save to serve man's purposes' (White 1967: 1205). Rather White's stress is on the kinds of technology introduced in the West in the early middle ages, and the ways in which this was underpinned by a particular view of Christian theology.

As has been already remarked, the Old Testament offers a positive view of the value of non-human life. Besides Psalm 104 and Proverbs 12 mentioned above, the divine speeches in the book of Job stress God's independent provision for wild goats, wild asses, ostriches, eagles and those fabulous creatures the behemoth and the leviathan (Job 39–41). Interestingly, Passmore disavows White's charges against both the Old Testament and Judaism, but persists in ascribing anthropocentrism to the New Testament, which he believes to have been influenced by Stoicism – hence his label of 'Graeco-Christian arrogance' (Passmore 1974: 15–17). Contrary to his view, New Testament references to birds and lilies in the Sermon on the Mount are not wholly anthropocentric; God is said to feed the birds, even though human beings are held to be of much greater value than them, and to clothe the lilies more richly than Solomon, implying a concern on the part of God independent of that for human beings (Mt. 6.26–30). Problem passages from the gospels, about the Gadarene swine (Mk. 5.11–14) and the cursing of a barren fig-tree (Mk. 11.12–14), are considered in the following section of this chapter, where Augustine's anthropocentric interpretations are contested. Meanwhile, the Apostle Paul, in whom Passmore detected the influence of Stoic anthropocentrism, included the whole of creation in God's plan of salvation (Rom. 8.21–23). So Passmore's double-barrel label, 'Graeco-Christian arrogance', appears to be doubly insecure, both with regard to the relevant Greeks (i.e. the Stoics) and also to Christians, or at least to those of the New Testament period.

White's contention that Christianity is more anthropocentric than other religions turns out to be similarly vulnerable. While some Christians have adopted anthropocentric views (Aquinas and Calvin being two notable examples; see Marlow 2009: 41–50), others have shown special concern for animals, for example Chrysostom and Basil, as discussed above. In case this might suggest an East-West division of attitudes, as Lynn White maintains, kindness to animals is also found in stories concerning Western saints such as the Northumbrian Saint Cuthbert (Waddell [1934] 1995: 52–62). Furthermore, the research of Susan Power Bratton and other scholars suggests that the supposedly large differences of attitude between eastern Greek and western Latin values were less absolute than White presumes (Whitney 1993: 163–4; compare Bratton 1988: 31–53). Indeed Bratton references detailed medieval studies showing that Francis of Assisi, whom White commends as a potential patron saint of ecologists, 'far from standing alone, is only one figure among a fully developed tradition of Christian appreciation of nature as God's Creation' (Bratton 1988).

Thus attributions of anthropocentrism to Aristotle, to the Stoics and to the Bible and Christianity turn out to have been exaggerated. But this finding does not of itself make any of these sources paradigms for environmentalists to draw on in the contemporary era. Before any such suggestion is considered, we need to reflect on what ancient sources convey about human responsibilities for nature.

Human responsibilities for nature

Our starting point for this inquiry is to consider the implications of Plato's stance in the *Timaeus*. Plato's holistic view of the world as a living creature (*Ti.* 30c) has indirectly influenced contemporary environmental action and policy in diverse ways. His societal and cosmic holism has served as an inspiration for various strands of environmentalism including that of James Lovelock, and J. Baird Callicott (Callicott 1980; Lovelock 1979). While Lovelock's Gaia theory probably owes little directly to Plato, and the name 'Gaia' was suggested to him by his classically educated friend William Golding (Lovelock 2006), Lovelock's views, which changed over time, well illustrate the variety of policy orientations that such a holistic stance can inspire.

In the earlier stages of his career, Lovelock stressed the view that the Earth can be compared to a superorganism, fully capable of self-repair. Hence, in his earlier work, he considers that environmentalist anxieties are exaggerations; having survived the explosion of Krakatoa, the biosphere is robust enough to survive the impacts of human technology, with little need for human intervention to preserve it (Lovelock 1979).[9] Subsequently, Lovelock became aware of the seriousness of global warming, and related phenomena such as the deforestation of tropical rainforests, and also of human interference with estuaries, which is in danger of halting the natural processes whereby bacteria produce vital chemicals such as dimethyl sulphide and methyl iodide. He therefore moves to advocating the importance of selective intervention to preserve or restore cosmic systems that are becoming overstressed (Lovelock 2006). Lovelock's later inclinations seem more closely based on empirical studies (including his own) of the contemporary disruption of global ecological processes. It is difficult to see how either of his positions is uniquely implied by cosmic or planetary holism.

Aldo Leopold's would-be application of eco-holism to ethics in his seminal work *A Sand Country Almanac* may serve as another example of holistic views. Here too, no single set of environmental principles or policies can be uniquely derived from this contemporary counterpart to ancient holistic ideas. According to Leopold: 'A thing is right when it tends to preserve the integrity, stability and beauty of the biotic community. It is wrong when it tends otherwise' (Leopold 1949: 224–5). While this principle may work for some areas of policy, such as biodiversity preservation, it cannot be used to underpin ethics in general, as appears to have been Leopold's intention. For breaking promises, telling lies and even acts of murder may turn out to be morally neutral or at worst even morally right, rather than wrong, when appraised against Leopold's criterion. So, although some environmental philosophers continue to be inspired by Leopold's land ethic (e.g. Callicott 1989), other bases appear needed for environmentalists to propound acceptable ethical principles. Likewise, the bearing on modern environmentalism of the passage of Plato's *Timaeus* that represents the universe as a living organism remains equally ambivalent (*Ti.* 30c), even if it has predisposed some people with a classical training to sympathy with the stances of eco-holists such as Lovelock and Leopold. The contribution of Plato's *Phaedrus* is arguably more positive and I will return to it below.

It might be argued that Aristotle would be a strong supporter of the pursuit of the biological sciences including the science of ecology, given his own investigations of biology and his attitude of wonder towards natural creatures. However when it comes to potential sympathy for environmental concerns, the evidence is that he was untroubled by the kinds of ecological change current in his day. One example of this is found in

Meteorology, where the context is the great cycle of nature, and the way in which various districts dry up for part of that cycle:

> Since there is necessarily some change in the whole world, but not in the way of coming into existence or perishing (for the universe is permanent), it must be, as we say that the same places are not forever moist through the presence of sea and rivers, nor forever dry. (*Mete.* I 352b16-20)

Aristotle's point here is that the universe, despite its fluctuations, is in the end qualitatively unchanging, and that the fluctuations eventually cancel each another out. This metaphysical assumption could well have been motivated by his acceptance of physical reality as a stable teleological world-system of entities reliably discharging their inbuilt purposes. But his view that the universe is permanent appears to pre-empt any concern about the planetary environment being fragile and vulnerable, except perhaps worries concerning temporary trends that could be expected not to last more than a few decades.

Thus even if, as argued above, Aristotle was a biocentrist and inclined to biophilia, there is little reason to regard him as an adherent of human responsibility to care for or preserve the natural environment. Indeed his very belief in nature's inbuilt purposes and the capacity of natural entities to discharge them suggests that there is no need for humanity to intervene to prevent, for example, deforestation or the growth of deserts (ecological problems already observable in his lifetime: see Hughes 1994: 5–6). Certainly some virtue-ethicists, such as Hursthouse, who largely follows Aristotle's account of the virtues in *Nicomachean Ethics*, regard themselves as Aristotelians (Hursthouse 1999). However their application of virtue ethics to attitudes to fellow-creatures is achieved through broadening the scope of the virtues that Aristotle commended, for example by including animals within the notion of kindness, rather than through adhering to the details of his own exposition (Hursthouse 2002).

The Stoic belief that everything is made for the sake of something else, except for the universe itself, is likewise unlikely to motivate contemporary societies towards adopting policies of preserving wild nature. According to Passmore, Stoics such as Seneca, in his *Natural Questions*, could look forward to future generations solving unsolved scientific problems and marvelling at the ignorance of his own day, but as long as everything could be understood as serving its natural purposes, not even concern for future generations would lead to concerns for nature preservation (Passmore 1970: 196). Present-day Stoic environmentalism, such as that espoused by Castelo and Holland, relies on making adjustments to the message of ancient Stoicism (Castelo 1996; Holland 1997), just as, *mutatis mutandis*, present-day Aristotelian environmentalism also does (Johnson 2005; Nussbaum 2006).

With regard to the Christian tradition, we have already noted Lynn White's claim that 'it is God's will that man exploit nature for his proper ends' (White 1967: 1205). His view presupposes that Christianity also harbours an instrumentalist attitude to the natural world, something that may be called into question. By contrast, Passmore's interpretation of biblical texts offers a different perspective, one that represents an improvement on that of White. As previously mentioned, Passmore exonerates the Old Testament and Judaism from fostering despotic attitudes to nature. Although he purports to find such attitudes in the New Testament, he also identifies what he considers to be a minority tradition within Christianity, in which humanity is God's steward or trustee of nature, responsible for its care. The development of this perspective, he suggests is most in evidence from the time of the seventeenth-century Chief Justice Matthew Hale. Previously Christian stewardship had been regarded as the stewardship of one's own talents or of treasure, rather than of the world of nature (Passmore 1974: 28–32).

Passmore cites Augustine to illustrate the supposedly despotic attitude of the New Testament, in which the passages about Jesus causing the deaths of the Gadarene swine and cursing a barren fig-tree are taken to indicate that refraining from killing animals and destroying trees is 'the height of superstition for, *judging that there are no common rights between us and the beasts and trees*, he sent the devils into the herd of swine and with a curse withered the tree on which he found no fruit' (Augustine, *Manichaean Ways of Life* 17.54, trans. Gallagher; see Passmore 1974: 111–12, his italics). But this passage tells us far more about Augustine and his polemical powers (against opponents such as the Manichaeans) than about Jesus. Stephen Clark suggests that the apparent narrative about the Gadarene swine may have originated in a parable (Clark 1977: 196), while Lk. 13.6–9 records a parable which could easily have been transformed into the story about the cursing of the fig-tree in Mark (Mk. 11.12–14).[10] Parables in the Synoptic Gospels, however, are widely held to have one single point of similarity with what they are introduced to illuminate, and are not to be regarded on the model of allegories, every item of which is expected to have a real-world correlate (see Dodd 1938: 12–26). Hence a parable about someone cursing a fig-tree for being unfruitful is unlikely to convey that Jesus or his followers do or should do likewise, provided that it had some other point such as that genuine followers will be like fruitful trees (Mt. 7.16). These problem passages, then, need not constitute obstacles to Jesus holding the biocentric attitudes shown in the passage about the birds of the air and the lilies of the field (Mt. 6.26–29), and suggested also by his consorting with the wild beasts in the wilderness (Mk. 1.13).

This being so, it is unclear whether Passmore is correct to suggest that the stewardship tradition, which recognizes that human beings have a responsibility for the care and protection of the natural world, was no more than a minority tradition within Christianity. We have already cited passages from Basil and Chrysostom that indicate care for non-human species carried over into the early centuries of Christianity (see further Glacken 1967: 192, 299–301). Susan Bratton relates numerous stories of the saints that imply this was a widespread stance during the first as well as the second millennium of Christianity (Bratton 1988). Furthermore, the care of the land and of its fertility by the first western monks in pursuance of the Benedictine Rule can reasonably be interpreted as 'Benedictine Stewardship' (Dubos 1974).

In its religious form, the stewardship tradition of human responsibility for nature is couched in terms of being answerable to God. However Passmore commends a secular version of it, clearly assuming that it could be held in the absence of this particular belief about accountability. In support of this he points to the passage from Plato's *Phaedrus*, cited above, which he rephrases as: 'it is everywhere the responsibility of the animate to look after the inanimate' (*Phdr.* 246b; see Passmore 1974: 28). This passage forms an important ancient source for belief in human stewardship of nature from an overtly secular philosophical work and carries no presuppositions about religious answerability. Secular versions of stewardship, for which the current generation of humanity is answerable for the care of nature not to God but to the community of moral agents, present and future, have been reiterated more recently by Richard Worrell and Michael Appleby, and also by Jennifer Welchman (Welchman 2012; Worrell and Appleby 2000).

Another promising ancient tradition is 'co-operation with nature', the view that the world was created incomplete, and that it is the role of humanity to enhance and/or perfect it. Passmore finds a version of this view expressed by the first-century BCE Stoic Posidonius, who wrote that the role of human beings included 'doing one's part in helping to establish the order [of all things]' (Clement of Alexandria, *Miscellanies* 2.21.129; Passmore 1974: 33). He also detects this tradition in the Hermetic text *Asclepius*, with its claim that 'God willed that the Universe should not be complete until man had done his part' (1974: 33), but takes the view that it disappeared from view from then until the Romantic writers of the late-eighteenth and early-nineteenth centuries such as Fichte and Hegel (1974: 33-4). But Glacken demonstrates that expressions of this attitude were also widespread among the Church Fathers, who used metaphors such as the world of nature being like a house, securely constructed but not furnished, the builder's plan being that perfecting it was the role of humanity. Examples of this, in addition to the works of Ambrose and Theodoret mentioned above, can be found in the writings of Basil the Great (Glacken 1967: 192-4), Augustine (Glacken 1967: 299-300) and Cosmas Indicopleustes, whose *Christian Topography* was written in the early sixth century CE (Glacken 1967: 300-1). So Passmore's view that the tradition of cooperation with nature was a minority tradition can be called into question, as can his parallel view of the notion of stewardship.

In commending the tradition of cooperation with nature, Passmore, an atheist, may well have envisaged a secular form of it, in which humanity, rather than completing the creator's handiwork, has the role of maintaining and enhancing the world of nature. At its best, this is a view that praises human creativity, both through art and architecture and through the construction of beautiful and sustainable landscapes. It also carries dangers, particularly if it morphs into a Promethean desire to take control of natural processes such as evolution, as advocated by the social ecologist Murray Bookchin (1991). While it does not necessarily lead to full-blown environmentalism, the tradition of cooperation with nature can feed into aspects of it such as the construction of botanical gardens and seed-banks, and also the creation of nature reserves and Sites of Special Scientific Interest.

Cooperation with nature may also reasonably be regarded as underpinning the Millennium Development Goals of 2000, and the Sustainable Development Goals of 2015.[11] In each case, United Nations conferences agreed that humanity as a whole needed to direct its efforts to the enhancement both of human life and society and of the biosphere, seeking to reduce or end starvation, to preserve biodiversity on land and in the seas and rivers, and to mitigate global warming. While these goals are less than perfect, they can readily be seen as salutary developments of ancient themes expressed by both pagan sources such as Plato and the Hermetic tract *Asclepius*, and equally by the early Fathers of the Christian Church.

Conclusion

Modern environmentalism has many sources other than the ancient roots discussed in this chapter. Significant among them are Darwinism and its implications; the development of the science of ecology; and the discovery that humanity, as well as being able to change the environment, is also vulnerable to it (Marsh [1864] 2003). While a few ancient writers such as Theophrastus were aware of man-made changes to local environments (Hargrove 1989: 25–6), the recognition of the general vulnerability of humanity and of nature, which surfaced in the nineteenth century and has gained momentum ever since, was probably a necessary condition for the birth of modern-day environmentalism, and is not explicitly present in texts from the ancient world.

Nevertheless, there were ancient practices that prefigured environmental concern, ranging from the biological study of species on the part of Aristotle and Theophrastus, and the vegetarianism of Neoplatonists like Porphyry, to the care of the land and of its fertility initiated by the followers of St Benedict and developed in earnest in the later middle ages by the Cistercians. Thus when René Dubos responded to Lynn White by suggesting the sixth-century St Benedict as the patron saint of ecologists in place of St Francis, there was a sound basis for his suggestion (Dubos 1974). Yet all these practices could not amount to environmentalism itself until the vulnerability of both humanity and the natural environment were understood, together with a sufficient understanding of natural cycles and of ways in which they were being disrupted, particularly in the epoch following the industrial revolution.

The ancient views on nature discussed in this chapter (and in the many other sources omitted here – tragedians, historians, epic and lyric poets and ancient writers on medicine) are thus insufficient to explain the rise of modern environmentalism. But arguably they are indispensable for its eventual emergence. Without the holism of Plato's *Timaeus*, some holistic strands of ecological thought would have been unthinkable, and the impetus might well have been lacking for the kind of systems theory that has proved pivotal to ecological science. Without the ancient views on human responsibility for nature traced by Passmore to Plato's *Phaedrus*, by Glacken to the Church Fathers and to the Old Testament, and by others to the New Testament, the stewardship approach might not have resurfaced in the seventeenth century, let alone been widely adopted in the twentieth and twenty-first centuries.[12] And without the tradition of 'co-operation with nature', traced by Passmore to the Stoic Posidonius and to the Hermetic tract *Asclepius*, and made both orthodox and commonplace by the Church Fathers, the belief that the role of humanity includes enhancing, as well as preserving, the natural world might well have failed to capture the imagination of the authors of the Brundtland Report on Sustainability (Brundtland 1987), the participants in the Rio Earth Summit (1992), and the authors of the Millennium Development Goals (United Nations 2000) and the Sustainable Development Goals (United Nations 2015).

If so, then the ancient origins of environmental thinking, in particular those aspects selected for discussion in this chapter, have been crucial to the rise of modern environmentalism. Whether modern environmentalism has itself acquired sufficient conviction and momentum to save the planet from the blights of global warming, pollution and biodiversity loss is another story, not to be investigated here.

CHAPTER 7
CREATURES IN CREATION: HUMAN PERCEPTIONS OF THE SEA IN THE HEBREW BIBLE IN ECOLOGICAL PERSPECTIVE
Rebecca Watson

Modern expressions of environmental concern often focus on the preservation and conservation of the natural world, yet this manner of relating to the Earth is understandably not typical of the Hebrew Bible.[1] This attitudinal gulf is nowhere more apparent than in respect of the sea; and indeed biblical modes of interaction with this realm often fall far outside contemporary paradigms. There is insufficient space in the present chapter to consider more than a fraction of Old Testament references to the sea, so instead it will focus on four areas, drawing attention to representative passages illustrating each. In particular, it is argued here that the view of the sea in the Hebrew Bible is more shaped by its perception of humanity as a creature within creation, as a small part of a much larger whole governed by God, than by the more-often discussed stewardship model of human relationship to the physical world.[2]

First of all, the writers of the Old Testament experienced fear and vulnerability in relation to the sea: they saw themselves as creatures caught up in forces greater than themselves and against which they could not protect themselves. Second, even the attempt to achieve mastery of the waves beyond a certain extent was a hubristic infringement of the limits of human attainment over against the divine. These two aspects therefore reveal a sense of human vulnerability and limitation. Third, marine life itself can excite a sense of wonder and awe, as well as an appreciation that dangerous beasts which might be inimical to human existence, and of no benefit to human needs, may still be of value to God and a source of pride to him. This, too, puts the role of humanity into perspective and denies it the centrality that has traditionally been claimed for it in Old Testament scholarship;[3] God is not anthropocentric and has concern for all his creatures. And finally, the Old Testament has a few scattered passages relating to fish, which acknowledge their defencelessness before fishermen and accept that the sea is not a limitless resource. Each of these four areas shall be discussed in turn with the aim of assessing how they may contribute to, or even challenge, contemporary thinking about the present ecological crisis.

Ecology and Theology in the Ancient World

Human vulnerability

In the Hebrew Bible, the sea is recognized as being entirely under divine rather than human control:[4] a strong sense of creaturely vulnerability,[5] fear of this dangerous aspect of God's creation[6] and awe at his mastery of it[7] precludes any thought of seeking to manipulate this environment.

There are many passages revealing this kind of perspective on the sea, but the portrayals of a storm at sea in Psalm 107 and Jon. 1.4–16 serve as good illustrative examples. According to Jon. 1.4–5, 'YHWH hurled a great wind upon the sea, and such a mighty storm came upon the sea that the ship threatened to break up. Then the mariners were afraid, and each cried to his god.'[8] The sailors instinctively appreciate that this is an act of God – and to be out at sea in such a storm is to be in a situation of profound human vulnerability, knowing that only a god could bring them into such danger and only a god could save them, and this is confirmed as the narrative develops. The captain wakes Jonah so that he, too, may call on his god, hoping that 'Perhaps the god will spare us a thought so that we do not perish' (Jon. 1.6). When this is to no avail, lots are cast to try to discern on whose account this storm has come upon them: it is imperative to discover the will and purpose of the god who has brought it about. The narrative makes very clear that they are in a sphere where they are totally reliant on the will of God, and that their efforts to lighten the load by throwing the cargo into the sea or attempting to row to shore cannot succeed. In preparing to throw Jonah overboard, the divine factor is uppermost in the sailors' minds: they are afraid that wrath may come upon them and pray, 'Please, O YHWH, we pray, do not let us perish on account of this man's life. Do not make us guilty of innocent blood; for you, O YHWH, have done as it pleased you' (Jon. 1.14). As soon as Jonah has been thrown overboard, the sea ceases from its raging, and the sailors then fear YHWH even more than they had before.

That a god is behind such a storm is already taken for granted in their responses and in the captain's words to Jonah, and indeed the sailors' reactions seem not merely to reflect the peculiarities of the Jonah story but to be consistent with reconstructions of the customary practices of mariners at the time (Brody 1998: 73–85). A very similar perspective is reflected too in Ps. 107.23–32, in which merchant ships are caught in a storm at sea. Again, it is explicit that YHWH 'commanded and raised the stormy wind, which lifted up the waves of the sea' (v. 25). The ship was violently tossed and the people were terrified, but

> Then they cried to the LORD in their trouble,
> and he brought them out from their distress;
> he made the storm be still,
> and the waves of the sea were hushed. (vv. 28–29)

In contrast to the passage from Jonah, there is no interest in Psalm 107 in the storm as a punishment or as serving any particular purpose, and the emphasis is on thanksgiving for deliverance rather than on why YHWH might have roused the storm in the first place. Here the raising of the storm seems to be viewed as 'the deeds of YHWH, his wondrous works in the deep' (v. 24). 'Deeds' here are *maʾăśê yhwh*, literally his 'doings', from the verb *ʿśh*, 'to do, make', but the 'wonderous works' are *niplāʾot*, from the root *plʾ*, 'wonder'. They are awe-inducing acts of power, beyond human comprehension, often not specified in their details but including miracles such as the deliverance from Egypt (Mic. 7.15, Ps. 106.22), the Jordan crossing (Josh. 3.5) and other deeds of salvation (Ps. 98.1). They also encompass all manner of divine action including just intervention in the human sphere (Job 5.9), the establishment and preservation of the creative order (Job 9.10) and other manifestations of power such as the sending of rain and snow (Job 37.5). They are signs of God at work. Thus, being caught out in a terrible storm at sea induces terror, so that 'their courage melted away in their calamity' (Ps. 107.26), but it also stimulates awe and wonder as well as profound gratitude for deliverance. Both of these passages show an innate sense of human weakness – the writers know what it is to be in the hands of God, utterly helpless and at his mercy in a world that is controlled by him and in which human endeavour is of no account. They are in a context in which huge forces are at work, vulnerable creatures in creation, who ultimately cannot even protect themselves let alone influence the forces of nature except through supplication to a benevolent God.

The implications of such a perspective for modern understandings of the sea are twofold. First, our safer and much more technologically advanced lives, although to some extent mitigating the harsher effects of nature, have arguably laid us open to a fallacy of self-determination and invincibility in the face of natural forces over which we still have no control. The sense of awe at a storm at sea is something in which we should share, even if we have not directly encountered such a phenomenon ourselves. The realization that human beings are part of nature, as creatures in creation, rather than somehow above it, is characteristic of the worldview of the Hebrew Bible and actually more prevalent than the notions of dominion to which many commentators instinctively turn. Passages such as Jon. 1.4–16 and Psalm 107 therefore invite a recentring of theological anthropology that places a greater emphasis on human weakness and dependence on God.

Potentially, another effect of understanding the sea (and indeed other great forces of nature) as beyond human control could be less positive: it could create the impression that nothing we do can alter or influence it. The environmental crisis of modern times, with rising sea levels, global warming, marine pollution and so on have clearly taught us that the sea is not immutable: it can be disturbed by human activity, and hence such thinking needs to be resisted. Nonetheless, the current situation has also shown us even more clearly our own powerlessness: we may inadvertently, out of ignorance and greed, have set in train mighty forces of change in the ocean, but it is not at all clear that we have the capacity to stop them. This, more than anything, teaches not just that we are vulnerable creatures within creation, but that our destiny is inextricable from that of the earth we inhabit, and that we are subject to great natural forces which we scarcely understand and which, when immediately faced with their threat, we cannot control. Undeniably, a sense of responsibility and restraint with respect to our relationship with the natural world is important, and indeed is already frequently abstracted from the *imago dei* language of

Gen. 1.26–27. However, the particular contribution of passages such as Psalm 107 and Jonah 1 to a contemporary theological anthropology is to offer a healthy counterbalance in their acknowledgment of human dependence and vulnerability, reminding us of the limitations of our understanding and capacity to rectify the harm we have caused or even to protect ourselves against natural forces that are so much greater than ourselves.

Human limitation

A second important aspect of the perception of the sea in the Hebrew Bible is exemplified by Genesis 2–3's definition of humanity (*hā'ādām*) as a creature of the ground (*hā'ădāmāh*), brought forth from the soil, set to till it, and destined to return there on death. The frequent delimitation of territorial boundaries as ending at the sea[9] further attests to a terracentric perspective, with the drama of the Red Sea crossing exemplifying the sense of the sea as (under normal circumstances) an impassable barrier.[10]

An obvious corollary of being a creature of the land is not to be a creature of the sea; hence, there may even be an extent to which to be on the sea is not quite to be in one's proper place in creation. Seen against this background, the sometimes negative reaction of the Hebrew Bible to the proficiency of a seafaring people such as the Phoenicians – who traded as far as Spain – is understandable. The Phoenicians' activities stimulated admiration and envy at their facility on the waves, the exotic locations they were able to reach and the opulence of the products in which they dealt. However, there is also a perception that such confidence in so awesome a divine realm is an act of hubris deserving judgment and humiliation. This is seen especially in the oracles against the Phoenician city of Tyre in Ezekiel 26–28.[11] It is not possible to examine these oracles in depth here, so instead a few central aspects of the condemnation of Tyre and their significance for the understanding of the sea will be highlighted. Perhaps the verses best encapsulating the grounds for the condemnation of Tyre are Ezek. 28.2–8, in which the Prince of Tyre is mocked for his pretensions to divinity and condemned to the underworld ('the Pit') and an ignominious death, without proper rites of burial:

> Because your heart is proud
> > and you have said, 'I am a god (*ēl*);[12]
> I sit in the seat/dwelling of the gods (*ĕlōhîm*),
> > in the heart of the seas,' (*ēl 'ānî/môšab ĕlōhîm yāšabtî bəlēb yammîm*)
> > yet you are but a mortal (*ādām*), and no god (*ēl*),
> > though you compare your mind with the mind of a god (*ĕlōhîm*) …
> They shall thrust you down to the Pit
> > and you shall die the death of the slain
> > > in the heart of the seas.

A key phrase here is 'heart of the sea(s)' (*lb[b] ym[ym]*), which performs an important role in these oracles, apparently communicating something of Tyre's essence and the nature of her claim to divinity. Probably the dominant aspect of the claim has to do with the location of Tyre, like the seat of God (or the gods), on the sea. The city's marine setting, however, has a dual significance in the present context: on one hand, it pertains to the geographic location of the city on an island;[13] on the other, it has a more immediate antecedent in the portrayal of Tyre as a merchant ship, richly fitted, furnished and manned from the products of her extraordinary trading enterprises.[14] Thus the motif of the heart of the sea(s) connects Tyre's maritime location and mercantile activity with aspirations to divinity and with the place of her death.[15]

However, there is a further dimension to the 'heart of the sea(s)' since a brief survey of occurrences of this phrase indicates that it is a place to which humanity does not belong. In fact, according to Ps. 46.3[2], the mountains (themselves understood as the abode of the divine) are in the heart of the sea.[16] The only named place or person specifically aspiring to be in the heart of the seas voluntarily, or to lay claim to them, in the Old Testament, is Tyre (most notably in 28.2, 'I sit in the seat of the gods, in the heart of the seas'; see also Ezek. 27.4, 'your borders are in the heart of the seas'); however, it becomes not just the place to which she ventures, heavily laden in her trade, but the place of her death: 27.25–27, 28.8.[17] As if to confirm that the 'heart of the sea' is not a region usually trespassed on by humanity, 2 Esd. 4.7 asks,

> How many dwellings are in the heart of the sea, or how many streams are at the source of the deep, or how many streams are above the firmament, or which are the exits of Hades, or which are the entrances of paradise?

This confirms the impression of the 'heart of the sea' as (like the other regions mentioned in this verse) effectively a divine realm, beyond human knowledge and understanding,[18] but it also suggests an association between the heart of the sea and the deep. There is a further connection between the various divine regions themselves (the heavens, underworld and paradise/Eden) on one hand and the heart of the sea/deep on the other. Particularly intriguingly, the passage from 2 Esdras alludes to dwellings in the heart of the sea, which seems to echo Ezek. 28.2, in which reference is made to the dwelling of God or the gods (*ĕlōhîm*) in the heart of the seas.

There are additional traces of the motif of the divine seat/abode in the heart of the sea in the Old Testament, as well as elsewhere in ancient Near Eastern sources. In the Ugaritic texts, El, the high god, is commonly described as having his dwelling 'at the source[s] of the rivers, amid the springs of the two deeps', (e.g. *KTU* 1.3.v.6–7, 1.4.iv.21–22).[19] A similar idea may be reflected in the river running out from under the door of the Temple in Ezekiel's own Temple vision (Ezek. 47.1–12; cf. Joel 4.18 [3.18]) and from Jerusalem in Zech. 14.8, as well as from Eden in Gen. 2.10.[20] It must be recalled, too, that the 'molten sea' (a vast, circular basin of water) stood in the Temple (1 Ki. 7.23–26; 2 Chr. 4.2–6), and this probably reflects the notion of a sea at (or under) the abode of God,[21] which may symbolically have been conceived as feeding the rivers which emanated from there. The statement in Ps. 29.10, 'YHWH sits/dwells at/over the *Mabbûl* [flood]'[22] could also express a similar notion. Although in this case, the allusion may pertain to the deity's heavenly abode,[23] parallels between heavenly and earthly Temple reflect the frequent lack of distinction between the two,[24] whilst the verbal idea of 'sitting' (*yšb*) provides a clear point of contact with Ezek. 28.2.

The concept of the sea as particular to the divine seems to have had wider currency in the Mediterranean world, since parallels may also be made with Greek thought. Sailing beyond the Pillars of Hercules (Straits of Gibraltar) was framed as an encroachment on divine territory,[25] and here too the sea was understood as having a sacred character as 'a boundary between the visible and invisible world, or between the world of humans, the gods, and the dead' (Beaulieu 2016: 16).[26] In 2 Esdras likewise the firmament, the underworld (itself close to the deep, as in Jonah 2 and Ezek. 28.8), and paradise (the garden) also all appear to be regarded as both a divine preserve and as located near the deep/sea. This seems to corroborate the evidence already assembled that all these regions have some form of association with the sea or deep, and are regarded as divine realms.

The reference to this peculiarly divine space in Ezekiel 28 is, of course, made clear in the mention of 'the seat of the gods in the heart of the seas' in verse 2. However, the same sphere of Temple imagery is adverted to again in verse 13, in the lamentation over the king of Tyre, in which is it said he was 'in Eden, the garden of God'. The association between Eden and Temple should occasion no surprise, given the sanctuary symbolism abounding in the portrayal of Eden both here and in Genesis 2–3, thereby identifying it as the archetypal abode of God and location of his presence (Wenham 1985). Within Ezekiel 28 specifically, the repeated emphasis in both parts of the oracle on wisdom (28.2–7, 12, 17) and beauty (28.7, 12, 17; cf. 27.3–4, 11) as the qualities of the Tyrian king permit an inference of his Edenic abode and divine status.[27] The connection between the Temple or abode of God and 'Eden, the garden of God' as the location for Tyre is further confirmed by the reccurrence of the same precious stones as a covering for Tyre in Eden (Ezek. 28.13, cf. Gen. 2.11–12) in the Temple, on the high Priest's breastplate (Exod. 28.17–19, 39.10–12).[28] The sin of Tyre, then, is its hubris in presuming to lay claim to the sea as its own, an attitude which was tantamount to claiming divine status and occupation of the garden or seat of God itself.

It seems reasonable to conclude, therefore, from such an understanding of the world, that it is not the proper role of humans to trespass with confidence onto the seas. Rather, excessive pride can only lead to destruction and is deserving of condemnation. This seems to be the message of the oracles against Tyre in Ezekiel. There is still much truth in the warning against the hubris of thinking human beings can possess and master the sea and make it their own. The fact that the surface of the moon is better mapped than the bottom of the sea is a striking reminder of our terrestriality, and of how much the heart of the sea is still an unknown and alien realm.

There is much more that could be said about the oracles against Tyre. However, for the purpose of examining human perceptions of the sea in the Hebrew Bible from an environmental perspective, account needs to be taken not just of the place of humanity vis-à-vis the sea, but also of biblical attitudes to fish and the other marine creatures. This will therefore comprise the focus of the second half of this paper.

The value of marine life

Despite the alien nature of the sea in Old Testament literature, nevertheless the diversity of life it supports can stimulate wonder, even at animals that may be dangerous or foreign to humanity. This is a cause for celebration of the purposes and creativity of God, as is seen in Ps. 104.25-26 and Job 40.25-41.26 [41.1-34]. However, such passages also reveal the mysterious nature of the creatures who were believed to inhabit the sea, since animals we would regard as mythological were understood to be part of God's created order.

Possibly the most striking aspect of Ps. 104.26 is the playfulness of Leviathan, a huge sea creature that elsewhere appears to have a draconic character.[29] The sea may be 'great and wide' (v. 25) and quite overwhelming to the human mind, but it is the perfect environment for this mighty beast, so much so that the sea itself almost seemed to invite his creation.[30] Playfulness also has a childlike quality that suggests that, for all his size and awesomeness to human eyes, he is a vulnerable creature, and that he is dependent on God for his life and sustenance. The reliance of all creatures on God for their preservation and indeed for life itself is brought out in the following verses (vv. 27-30), suggesting that this may indeed be the train of thought that the psalm pursues here in respect of Leviathan.[31]

The most detailed passage to portray this beast, Job 40.25-41.26 [41.1-34], depicts him as a truly terrible animal, invincible in his strength and in the impenetrability of his scales and other awesome physical attributes.[32] Of particular interest, notwithstanding the terrible form of Leviathan, is the perspective of the author on this animal, which is to imagine God's pride and delight in this creation (41.4[12])[33] before which even the divine beings are not safe, and who is impossible to tame, capture or kill, yet which on earth has no equal[34] (41.25[33]) and is 'king over – or above – all the majestic creatures' (v. 26[34]).[35] This animal lives in an alien environment to humanity and, far from being conducive to human wellbeing, actually poses a potential threat to it, yet nonetheless is not only a creature of YHWH, but YHWH himself revels in its invincible strength and the fear it can cause even among the heavenly beings. This passage therefore constitutes a celebration of this awesome creature as part of God's paradoxical and incomprehensible world.

Thus Psalm 104 and Job 40.25-41.26 represent contrasting pictures: on one hand, a depiction of an animal at home and relaxed, undisturbed in its natural environment, and on the other hand, a picture of conflict, in which Leviathan resists human efforts at restraint, capture and slaughter. Between Psalm 104 and Job 40.25-41.26, therefore, Leviathan is revealed as both an animal of power (see especially Job 41.4-17, 22-26 [41.12-25, 30-34]) and of creaturely dependence (Ps. 104.24-30), of playfulness (Psalm 104, cf. Job 40.29 [41.5]) and of ferocity (notably in Job 40.25-41.3, 17-21 [41.1-11, 25-29]), one who is king over the sons of pride (Job 41.26[34]), fearsome (Job 40.32-41.3, 17 [41.8-11,25]) and invincible (40.25-41.26 [41.1-34], yet also nurtured by God (particularly in Ps. 104.24, 27-28, 30; cf. Job 41.25[33]) and a source of divine delight (Job 41.4[12], cf. Ps. 104.26).[36] Insights that are of especial contemporary relevance are the appreciation of the intrinsic value to God of creatures that are apparently irrelevant or even threatening to human existence, and this presents a strong counterweight to narrow self-interest and utilitarianism. However, these passages also speak of these animals' innate vulnerability. Their dependence on God for their sustenance and for life itself might reasonably be interpreted in practical terms as an utter reliance for their continued existence on the order that he has created and which must be sustained if such life should continue. The contrast between the irenic Leviathan frolicking in the ocean and the fierce monster resisting human efforts at mastery ('[T]hink of the battle; you will not do it again!', Job 40.32 [41.8]) possibly also has resonances for situations where the conservation of predators encounters the difficulty of animals being constrained by human encroachment on their habitat, leading to conflict and vulnerabilities on both sides. In particular, human infringement may reveal a very different aspect of an animal than when it is seen free and unconstrained within its natural environment.

The limitations of fishing

However, the Old Testament has more to say about aquatic life than simply to wonder at it, and this leads into the fourth aspect of biblical thinking about the sea that shall be explored here.

There are a number of passages in the Old Testament that, despite viewing fish as a source of food, also reveal some concern for the fish themselves, whilst Num. 11.22 shows an appreciation that they are a finite resource. Here, YHWH has just promised to give the people plenteous meat in the wilderness for a whole month 'until it comes out of your nostrils and becomes loathsome to you' (v. 20) but Moses' response is incredulity: there are six hundred thousand people.[37] 'Are there enough flocks and herds to be slaughtered for them? Are there enough fish in the sea to be caught for them?' he asks rhetorically (v. 22), implying a strong denial: of course there is not sufficient to feed that number of people.[38]

From the perspective of the meeting of ancient and modern cultures through this verse, the words of Moses, in their disjunction with attitudes which have been prevalent for too long and which are still pervasive today, present a challenge to the fallacy of assuming that the sea has limitless capacity to give. With primitive fishing methods and relatively small populations, it may have been possible to take all that one was able and for fish stocks to remain buoyant.[39] However, with modern fishing techniques, the recognition that there are not enough fish in the sea to feed everyone is vital if they are to be harvested sustainably, rather than the population of key species being allowed to collapse.

In addition, the words of Moses may reflect a different attitude to the fish than is prevalent in our own culture. Many modern translations read (felicitously in English, but diverging from the Hebrew), 'Are there enough fish in the sea to catch[40] for them?',[41] using an active verb to place the emphasis on the human need and the action of catching. Others go further in putting humanity in the centre of the action: NIV, for example, reads:

'Would they have enough if all the fish in the sea were caught for them?'[42] However, the Hebrew places the focus on the fish, who (in the passive) will be 'gathered' (i.e. in a net and removed). As such the attention is drawn to the fish and what is done to them, rather than to the direct action in addressing the need for meat. Translated literally, the Hebrew asks, 'Shall all the fish of the sea be gathered for them and there will be enough for them?' This could be understood in the conventional way, as implying that even if all the fish in the sea were caught, there would not be sufficient to feed the multitude.[43] However, it could alternatively be construed as asking, 'Should such a thing be done?' Shall all their cattle and flocks be killed so that there is enough meat for them to eat? Should all the fish of the sea be caught so that there are enough to eat? In this case, a shocking prospect is presented, since to kill all the livestock would be to cut off all future sources of milk and meat, and to catch all the fish would have a similar effect. Feeding so many would necessitate such radical action but of course it cannot and should not be done, not least because it would be a short-sighted step to the annihilation of the people themselves, but possibly also in taking some account of the animals concerned as well. A further aspect of this, which seems to be innate to the perspective advanced in the pericope, is the sense of excess in the demand for meat, and the idea that a month's supply would be enough to cause revulsion; hence, such a thing should not be done also on moral grounds, because such a strong craving is of itself ethically wrong, and the daily consumption of meat is unnecessary and inappropriate.

In addition to the appreciation of the finitude of the fish of the sea, there are also a few passages that present the plight of people suffering from invading forces under the metaphor of fishing. This is important because it reveals an ability to relate to the experience of the fish, even as fishing was an important means of subsistence. Two examples will be considered here.[44]

Ecclesiates 9.12 likens human beings being snared at the 'evil time',[45] which will suddenly fall upon them, to fish being 'caught in an evil net' or birds being held in a snare. The point here, as the preceding words state, is that 'no-one can anticipate the time [of disaster]'.[46] The verb applied to the fish, *ʾāḥaz*,[47] denotes the idea of being seized or caught, with the added concept of being taken possession of. They are grasped and held in the net, then taken away by the hunter. Particularly striking here is the use of the same adjective, *rāʾāh*, 'bad, evil', to describe both the occasion when humanity is snared (the 'time of calamity') and the net in which the fish are caught (the 'cruel net'),[48] a choice which implies a value-judgment about fishing as seen from the perspective of the victim. Clearly, the analogy is used simply to highlight the horror of the time that is to come, but this nonetheless reveals a capacity for relating to the plight of the hunted animal, without which the metaphor could not have rung true to the audience.[49]

The vulnerability of the fish is conceived in a slightly different form in Hab. 1.14–17. Here they are described as having no ruler (*mōšēl*), implying that they lack direction and protection and thus can put up no defence against the enemy with his rod and net[50] who appears in the next verse. His rejoicing and exulting at his success is clearly felt to be odious, and the question, 'Will he therefore keep emptying his net and continue to kill the nations without pity?'[51] implies that there is a proper limit to what should be taken.

This misguided and ugly greed is caricatured in the picture of the foe sacrificing to his net and seine since he has gained so much by them and evidently seeks for more. Although this constitutes a metaphorical portrayal of the fate of the mass of humanity[52] before a great conqueror who is intent on destroying one nation after another, nonetheless, the passage illustrates a capacity for perspective-taking which perceives the ugliness of excessive greed as it may be manifested in the exploits of a rapacious fisherman and appreciates the innate vulnerability of aquatic life. In the acknowledgement of the finitude of fish as a food source and in the recognition of fish as vulnerable creatures in the face of human rapacity, there is much with which we can concur, not least as having now reached a point of serious depletion of fish-stocks through greed analogous to that which is here condemned.

The characterization of the fishes' predicament as the lack of a *mōšēl*,[53] a ruler who may lead and protect them,[54] speaks both of human and piscine need.[55] The tenor of this metaphor speaks of divine absence for the human victims of Babylonian expansionism and possibly also of political turmoil within Judah itself.[56] But how might the problem of the absence of a 'ruler' for the fish, and their consequent vulnerability in the face of human rapacity, be addressed? The same verb, *mšl*, recurs in Ps. 8:7[6], 'You have given them dominion [more literally, "rule"] over the works of your hands', so this passage may appear to offer the tantalizing prospect of a solution in the ideal of 'stewardship'[57] by a humanity who recognizes itself as one creature among others and with responsibility to ensure the welfare of all.

Nonetheless, this suggestion needs to be counterbalanced by the reticence of the Bible as a whole to make rulership claims for humanity – or at least to envisage this as being worked out in the grand manner modern writers might imagine – and by its portrayal of the animal kingdom as directly dependent on God (as in Ps. 104), as well as at times vulnerable before human activity.[58] These biblical ideas encourage a greater awareness, first, of the inimical effects of human action; second, of the incompatibility of human threats to marine fauna with passages affirming God's creation, blessing and preservation of a range of animal life; and, third, humility before making claims to have the wisdom or capacity to enact a God-like regency over creation, despite the urgency of the need to learn to act in accordance with what the Bible presents as his will for it. God's provision for his creatures is elsewhere celebrated as a role peculiar to the divine rather than one devolved to humanity, and indeed it is invoked on such a grand scale as to be well beyond human understanding and capacity.[59] The dependence of humanity itself on God, and their extreme vulnerability without the benefit of his direct protection and 'rule', is of course, the concern of Hab. 1.14–17 itself.[60]

Conclusion

In conclusion, it may be argued that the Old Testament sense of human finitude before creation is one that urgently needs to be recaptured, not least where we have tampered with what Job refers to as 'what I did not understand, things too wonderful for me, which

I did not know' (Job 42.3). The perception that hubris in believing in human mastery of the seas leads to destruction accurately expresses a lesson that we are only now beginning to learn. At the same time, the Hebrew Bible is able to express wonder at creation for its own sake, freed from utilitarian illusions and even going so far as to celebrate those creatures which may potentially be detrimental to human safety (Psalm 104). This sense of creatureliness in creation and co-existence rather than mastery is one that provides a model that, though often outside the modern horizon, is increasingly recognized as necessary for ecological flourishing. Finally, the finitude of fish resources is prescient for the need to limit human exploitation after generations of misguidedly presuming on the limitlessness of the sea. The big challenge is whether it is possible, after centuries of scientific and technological development, to recapture a 'primitive' worldview that dispenses with the illusions of control and allows us to recognize vulnerabilities which have long been denied but which have been as much exacerbated as mitigated by generations of human endeavour.

CHAPTER 8
RECONSIDERING THE CHTHONIC IN AESCHYLUS' *ORESTEIA*: ERINYES, THE EARTH'S RESOURCES AND THE COSMIC ORDER

Emmanuela Bakola

Introduction: Eco-theology and the 'chthonic'?

There is an ancient category of the divine which would seem *sine qua non* in a volume about the relationship between ancient and modern ecotheological thinking: that of the 'chthonic'. As the term (from the Greek *chthon*=earth) suggests, these divine powers' sphere of operation is the earth, as the source of fertility and growth.[1] The Athenian cult and festivals of deities such as Zeus Meilichios (Diasia) or Polieus (Dipolieia), Dionysus (Lenaia, Rural Dionysia, Anthesteria, Oschophoria), Demeter and Persephone (Proerosia, Thesmophoria, Haloa, Great Mysteries) – to name but a few – reflect to a large degree the human need to establish a good relationship with powers which facilitate or hinder the natural processes of production, generation and growth; the powers, that is, that can bestow prosperity, livelihood and wellbeing.[2] Such cult and festivals reflect an awareness that these processes, whilst outside human control, can be influenced by human behaviour towards the divine and towards nature. In other words, if we are to look for an 'ecological' awareness in the various theologies of the Greek world, a fruitful area would undoubtedly be the category of divine known as 'chthonic' and associated cults.

Unfortunately, due to a long-standing scholarly debate on the antithesis 'Olympian vs chthonic' in Greek religion,[3] the term 'chthonic' has been more frequently understood as reflecting the dark qualities of the underworld and the dead. It is not uncommon to find discussions of the category leaning heavily towards this definition, with mere lip-service paid to these deities' more 'positive' aspects, namely concern with the earth's generative processes, with growth, fertility and well-being. Often, when these aspects are discussed, they are understood only in connection with agriculture, not more generally with spontaneous growth, natural production and the conditions which allow them.[4] As a result, 'chthonic' has become too narrow a concept, and its perceived negative aspects and associations with 'primitive' elements of Greek religion have only contributed to its suppression in scholarly discourse. This is unfortunate, because the chthonic has a prominent place in hugely popular cultural forms, especially Greek drama. Furthermore, in the last thirty years, the debate about the chthonic has become even more rigid because of the scholarly focus on types of sacrificial ritual, and the rejection altogether of a single 'chthonic' type of sacrifice.[5] All these difficulties have progressively resulted in a reluctance even to employ the term. For example, although Parker analyses the category of powers who promote growth and are associated with death, in the same publications he has also expressed concerns over the usefulness of the term 'chthonic' as analytical tool.[6]

Parker's reluctance largely stems from the observation that the Greeks did not use the term *chthonios* consistently to describe these powers.[7] Although this is instructive, absence or inconsistent use of a term does not amount to absence of a concept. Since, as even Parker's analyses show, a group of divinities have been consistently understood to share a set of powers or associations, instead of doing away with the modern term 'chthonic', it would be more productive to try to understand what function the term and concept have been serving for so long. Unfortunately, different scholars have defined both 'chthonic' and associated terms very differently.[8] Above all, not every scholar understands the concept 'earth' and its properties in the same way: as mentioned above, the dimension of growth (not only agricultural, i.e. engineered, but also spontaneous growth) does not receive equal attention in most scholarly engagements with the chthonic. This is mainly because influential scholars misrepresented *chthon* as reflecting the cold earth of the underworld, a realm of decay and death. Perception has largely been shaped by Chantraine's entry in the *Dictionnaire Etymologique* s.v. *chthon*, heavily influenced in turn by Wilamowitz-Moellendorff's creation of a dichotomy between *chthon* and *ge* – the former as the 'cold earth' that houses the dead, and the latter as the source of production[9] – despite multiple passages (already in *LSJ*) showing that this dichotomy is false.[10]

This chapter sets out to redress the balance regarding the modern definition and understanding of the chthonic. It also intends to show that its narrow perception as a technical term and in opposition to the 'Olympian' may be hindering us from accessing Greek ecological attitudes and their place in certain philosophical discourses that emerge in archaic and classical sources. There are, naturally, a number of questions and problems which may have discouraged the exploration of the term in this manner until now: first of all, the question of whether our sources support an approach to the chthonic that puts equal weight on both sides of the earth, the productive and the deadly. For, while divine entities like Demeter, Persephone, Pluto(n), Hades and Hecate may be linked equally to the world of the dead and the productive powers of its depths in many of our sources,[11] would this polarity be applicable to *all* divinities connected with the earth, including the terrifying divine agents of the underworld?[12] Furthermore, when we talk about powers of generation and growth as key to the definition of the chthonic, do we mean *only* the powers of the earth below? What is the relationship of the chthonic to other natural elements, spaces and associated divinities that are acknowledged to have similar powers? And ultimately, is 'chthonic' as analytical category suitable for the modern study of Greek literature and culture, or would its use require a return to the nineteenth- and early twentieth-century approaches to fertility rites?

Reconsidering the Chthonic in Aeschylus' *Oresteia*

Answering these questions requires more space than this chapter allows, so I do not aim to be exhaustive. Starting from the latter question, on the term's applicability for the study of Greek culture, the recent resurgence in scholarly study of Hesiod's epics *Theogony* and *Works and Days* (eighth or seventh century BCE) and their profound influence on the philosophy and literature of archaic and classical times is enormously useful.[13] Both Hesiodic works provide excellent glimpses into Greek concepts about the earth, and *Works and Days* especially focuses on humans' relationship to the soil, on understanding the signs of nature, conducting labour according to them, and using earth's resources carefully. It envisages knowledge of the land and respect for the rhythms of nature both as a prerequisite for ensuring productivity and as a fundamental form of justice and cosmic order.[14] Crucially, *Works and Days* 111–26 and 248–55 connects the natural and cosmic order with chthonic powers that bestow wealth to the just and deprive the unjust from means of livelihood.[15] *Works and Days* is a telling example of a work of literature where theological, philosophical and cosmological approaches to the earth converge and overlap. I believe that this combined approach can do far more justice to the concept of the chthonic than its treatment merely as a technical term of Greek religion.

In this chapter on Greek ecotheological attitudes, I argue that the chthonic may also be understood from a combined theological, philosophical and cosmological perspective, which concerns the relationship of humans to the soil, the earth and its generative powers. I will construct this argument in relation to the *Oresteia* trilogy of the Greek playwright Aeschylus, a work that was deeply influenced by Hesiod.[16] The *Oresteia*, a 'treasure house of chthonian concepts' as it has been appositely characterized,[17] provides previously unappreciated evidence that the chthonic captures a certain consciousness related to natural generation and production, the use and abuse of resources (the 'wealth of the earth') and human relationship to the cosmic order as determined by these. By exploring how earth and the chthonic inform the role of the Erinyes in the trilogy and demonstrating that their function goes well beyond that of deities of the primitive order of revenge and their antithesis with the Olympian gods, I will demonstrate that their theology, in fact, approximates an ecotheology. I will also explore how this category of the chthonic relates to bigger cultural questions, including the relationship of humans to the cosmic order, which is a question at the heart of the *Oresteia*.

Beyond the 'Olympian vs chthonic' antithesis: A renewed focus on the chthonic per se in the *Oresteia*

The trilogy of Aeschylus known as the *Oresteia* (458 BCE), which comprises the plays *Agamemnon*, *Libation Bearers* and *Eumenides*, is a breathtakingly complex poetic text of nearly 4,000 lines. The story line is deceptively simple, and at first does not seem to relate to the ecotheological concerns raised above: in *Agamemnon*, following the military expedition against Troy and the sack of the city, the eponymous hero returns to his palace in Argos. His wife Clytemnestra receives him (in the pivotal 'tapestry scene', for which see below) and then kills him in revenge for the sacrifice of their daughter Iphigeneia, which he had perpetrated ten years earlier to make the expedition to Troy possible. In the second play, the *Libation Bearers,* Clytemnestra's son Orestes returns to the palace from exile and kills his mother to avenge his father's murder. He does so both following divine orders by the Olympian god Apollo and under the threat of the Erinyes, deities of the earth who avenge injustices. After the matricide, Orestes is hunted by the Erinyes until, in the third play, the *Eumenides,* he takes refuge in Apollo's oracle at Delphi. There, he secures the support of the god, who assumes responsibility for ordering the matricide. The Erinyes are forcibly ejected from the oracle by Apollo, and Orestes is told to escape to Athens, where he is finally tried in a court of law, the Athenian Areopagus. He is acquitted of matricidal guilt and allowed to return home. However, the trilogy does not end here. It closes with a more unexpected and, apparently, more significant homecoming than that of Orestes: the homecoming of the Erinyes, who are persuaded by the goddess Athena to accept the Athenian earth as their new home. In a final celebratory procession, they are given the responsibility for overseeing trials of murder and protecting the life, fertility and productivity of the Athenian land and its people.

The *Oresteia* is one of the most exhaustively studied works of Greek antiquity, with studies varying hugely according to their interpretative focus.[18] In different periods, readings have tended to focus on different areas, and it is a long time since theological concerns and the chthonic, in particular, have been at the centre of scholarly interest in the trilogy. Making sense of this dimension of the *Oresteia* was always dominated by the contrast 'Olympian vs chthonic', namely the contrast in the divine world of the play between the new (predominantly male) and the old (predominantly female) divine order, the former including gods like Zeus, Apollo, Athena and Artemis and the latter the Erinyes.[19] This contrast takes centre stage in the *Eumenides,* after the Erinyes clash with Apollo over Orestes' matricidal guilt (*Eum.* 179–234 and 574–753). Here, the chthonic order, which represents the old system of retributive justice, seeking the punishment of every violation of the cosmic order with more violence, is shown as defeated by the order of the new gods and the justice of law, as Orestes is acquitted by the Areopagus court and the goddess Athena. However, with Athena's intervention, the Erinyes accept their new honours in the land of Athens and are installed in its underground recesses as guardians of the newly instituted Areopagus (*Eum.* 794–end). Thus the chthonic deities willingly accept their position in the new order, because one element of the previous system of violent revenge with which they were associated, namely fear of punishment, remains necessary for social order.

Whilst the stark contrast between Olympian and chthonic is quite prominent in the *Eumenides*, it does not feature as such in earlier parts of the trilogy. On the contrary, as several scholars have pointed out, the divinities of the Olympian and those of the chthonic order are shown from the start as working in parallel.[20] Furthermore, all major gods of the trilogy – Zeus, Apollo, Athena, Artemis – have themselves prominent chthonic aspects and links to the earth. The opening scene of the *Eumenides*, for example, goes to great lengths to stress the Delphic Apollo's ancestral lineage from earth deities (*Eum.* 1–14), and it has correctly been argued that Athena's portrayal in the *Eumenides* draws on her cult as Athena Polias, the version of the goddess that has strong links to chthonic cult.[21] It has also rightly been noted that the sacrifice that Artemis demands for her appeasement, the sacrifice of Iphigeneia, also has characteristics of chthonic cult that are hard to ignore, including the fact that the goddess had earlier received a 'sacrifice' of a pregnant victim.[22] Furthermore, Zeus' connections with subterranean justice and the Erinyes are flagged from the start of the trilogy and pervade it to the end.[23] Therefore, the fact that the trilogy as a whole closes with the acceptance of the chthonic into the divine order of the Olympian gods is not a sudden development, as is often thought, but built solidly on earlier representations. This suggests that the Olympian vs chthonic antithesis is far from simple and that defining chthonic as the opposite, the 'other', of the Olympian does not do justice to this category. Even more so, the fact that the ending of the trilogy chooses to focus on the Erinyes beyond any other divine or human agent suggests that the role and meaning of the chthonic per se are key in the *Oresteia*.

Elsewhere, the theatricality of the trilogy corroborates this observation, as it constructs an almost continuous presence of the chthonic in both human and divine worlds in a striking and multifaceted manner. The *Libation Bearers*, in particular, teems with the presence of the earth: more than half of the play's length focuses dramaturgically on the ground and the tomb of Agamemnon, with the characters kneeling, touching, examining and beating the earth, and calling upon the infernal powers (*Cho*. 1–509), appearing as if drawing on the earth's wrath, succour and energy. Scenes such as the description of the dream sent by the earth's powers (*Cho*. 32–54), the pouring of libations into the earth (*Cho*. 84–166), the focus on Orestes' and Electra's footprints on the ground (*Cho*. 204–10) and the anticipation of dead Agamemnon's epiphany from his tomb (*Cho*. 455–509) make the dramaturgy of the play rich with an intense presence of the earth. Much of the effect of the *Libation Bearers* echoes pivotal moments from the earlier *Agamemnon*, especially the 'tapestry scene', where the focus on the ground and on nature's productive powers is striking (*Ag*. 906–74). The chthonic, 'that of the earth', is dramaturgically shown to be much bigger and more multifaceted than a mere embodiment of the old system of justice, or a foil to the Olympian order.

In the last twenty years, scholars such as Pat Easterling and Helen Bacon have argued that the Erinyes have a more central role in the *Oresteia* than has generally been thought, corroborating earlier arguments by Oliver Taplin, R. P. Winnington-Ingram and Ruth Padel.[24] Recently, I have shown that the Erinyes have a weighty presence on stage as embodied presences: in particular, I have argued that the servant women who emerge from the house of the Atreids in the *Agamemnon*'s tapestry scene, and who dominate the *Libation Bearers* as the play's chorus ferociously advocating the matricide, are identified with the spectral presence of the chthonic deities said to be permanently lurking in the house.[25] I have also demonstrated that, as elsewhere in Greek drama, the space of the house in the *Oresteia* has a deeply symbolic function and captures both sides of the chthonic, natural productivity and wealth as well as death.[26] This is especially striking in the pivotal 'tapestry scene' (for which see below).

These developments provide a solid foundation for revisiting the role of the chthonic in the trilogy, especially in its final scene, where the Erinyes, deities of the earth, are installed in the Athenian soil as guarantors of its natural generative processes and the people's prosperity. The contrast between the Erinyes here and the Erinyes as they appeared earlier in the trilogy – fierce punishers of hubris, implacable hunters of killers, *daimones* who induce madness in the minds of their victims – appears enormous and demands an explanation. For, this is the scene which 'brings not merely a play, but a vast and complex trilogy to completion', as Winnington-Ingram stressed.[27] Examining the chthonic in a much wider scope is thus important not just for the sake of understanding the Erinyes and their theology, but in order to gain a deeper insight into the *Oresteia* as a whole, especially its engagement with thinking about the earth and how this is connected, more widely, with ideas about the cosmic order.

The 'transformation' of the Erinyes and the role of the chthonic at the end of the *Oresteia*

Until v. 777 of the third play, very little suggests that the Erinyes share anything with divinities that are concerned with wealth, prosperity, fertility, the preservation of life and the natural generative processes. Yet, from 778 onwards, these very ideas become central to the negotiations between the Olympian and the chthonic order, and instrumental to the climax and closure of the entire trilogy. Close to half the length of the final scene (115 verses in total) focuses on them, starting at *Eum.* 778. As these passages have received disproportionately little attention from scholarship, I will offer an analytical summary, in order to show the centrality of the themes in question.

The acquittal of Orestes by the Athenian court of the Areopagus is devastating for the Erinyes; in revenge, they threaten to release poison that will bring disease to the land of Athens, making it sterile and childless (*Eum.* 777–92). Athena tries to convince the deities that they were not defeated, begs them not to make the land sterile, and reassures them that the land of Athens can be their new home, where, if they stay, they will receive official cult worship (*Eum.* 800–07). The Erinyes repeat their threats (*Eum.* 808–22), but Athena argues that if they are persuaded, they will receive the opposite of what they threaten: sacrifices for protecting crops, childbirth and marriage (*Eum.* 824–36, 887–91). At *Eum.* 892–901, the Erinyes seem to soften their anger and pay attention to the goddess, and Athena once again reassures them that their cult will be linked to the prosperity of every Athenian house (*Eum.* 892–901). She explains that the Erinyes' blessings will come from every aspect of the natural environment: from the land, from the waters of the sea, from heaven, from winds and from sunshine; they will guarantee fertility for the Athenian people, their livestock and their crops (*Eum.* 901–13).

Finally, at *Eum.* 916–26 the Erinyes accept Athena's offer to take residence with her, Zeus and Ares, protecting the city and making various natural blessings burst in profusion from the earth. At *Eum.* 938–48, the Erinyes confirm that they do as Athena asks: no unfavourable winds or diseases will destroy their crops; the animals will be cared for and have multiple births at the appropriate time; the next generation of citizens will enjoy an equally blessed land. At *Eum.* 956–67, the Erinyes assert that they will prevent early deaths of men and ask other gods to bless women with marriages and at *Eum.* 976–87 they confirm that they will prevent internecine slaughter. From *Eum.* 996, where the deities start moving towards their new home under the earth, the trilogy is reaching its final, climactic moments. The Erinyes praise the Athenians for the wealth that they gained justly (*Eum.* 996–1002). Athena leads the procession and asks the goddesses to send up good things from their underground abodes and the citizens to be grateful for the blessings (*Eum.* 1003–13). The Erinyes express thanks and reiterate that their new habitation will bring blessings to Athens (*Eum.* 1013–20). Finally, at *Eum.* 1033–47, the processional escort send the Erinyes off as they enter their new home in the depths of the Athenian earth.

This striking 'transformation' of the Erinyes at the end of the *Oresteia* has been discussed by several scholars, with the concept of the chthonic important in its interpretation.[28] The most widely accepted and most economical approach suggests that we do not have a 'transformation' in the literal sense of the word, but a manifestation of the opposite side of chthonic deities in Greek religious imagination; since decay/death and fertility/life are the two sides of the earth, the benevolent aspect of a chthonic power is to bestow fertility, prosperity and wellbeing; once the Erinyes' anger and readiness to inflict punishment are appeased, the deities manifest themselves in the 'positive' version of the chthonic.[29]

This is, to my knowledge, the only instance in Greek scholarship where the Erinyes, *qua* chthonic deities, are considered to have benevolent qualities connecting them with wealth, natural productivity and the generation of life. Here even chthonic forces so closely associated with deadly curses, terrifying violence and revenge, have a role in the renewal of life and nature as well as the bestowal of prosperity and wealth. Unfortunately, this 'anomaly' has largely been resolved by recourse to the theory that the Erinyes' identification with benevolent forces is an invention by the playwright and has little, if anything, to do with the divinities' role earlier on in the trilogy.[30]

However, none of the interpretations offered so far explains why Aeschylus made the Erinyes' manifestation as earthly deities of fertility and life the climax and closure of the *Oresteia*. If we assume that the benevolent side of the Erinyes' chthonic nature is activated only in order to align them with other developments of a political or a social nature (such as the institution of Areopagus and the end of the vendetta code), we implicitly accept that the main component of the ending has a loose connection with the rest of the *Oresteia*.[31] If, however, we are not satisfied with this assumption, then the natural conclusion is that the 'new' role of the Erinyes must be meaningful for, and organically connected with, the rest of the trilogy. Exploring this prospect is timely now that it has been demonstrated that the Erinyes have a much more integrated and continuous presence in the trilogy – due to their regular apparitions at key moments throughout – than previously thought.[32]

The Erinyes and the economic order in the *Oresteia*

At *Ag.* 146–55 we encounter the first of many passages that refer to the sacrifice of Iphigeneia (*Ag.* 154,), undertaken to allow the imperialist expedition against Troy to set off. In this key passage, the Erinyes' preoccupation with avenging a violent death has a pronounced *economic* dimension. Whilst the Erinys (singular of Erinyes) of the passage is referred to as *Menis,* 'Wrath', 'Fury', and is said to be waiting guilefully to avenge the sacrifice of Agamemnon's young daughter,[33] in the same context, she is also called *oikonomos,* 'manager of the house's wealth':

> for there awaits, to arise hereafter, a fearsome, guileful *oikonomos,* a Wrath that remembers and will avenge a child. (*Ag.* 154–5; trans. Sommerstein 2008, adapted)[34]

The intriguing suggestion is that the murdered child was the house's wealth, for which the Erinys, as *oikonomos,* was responsible. This puzzling connection between the Erinys and *wealth* becomes somewhat clearer as the trilogy progresses, and as the Erinyes are repeatedly envisaged in association with economic order, and as avengers of its disturbances.

The first two songs (choral odes) that the chorus performs in *Agamemnon* are key for this development. Almost 200 verses after the reference to Erinys-*oikonomos,* in the first ode (*Ag.* 355–487), it is suggested that Paris', Agamemnon's and Menelaus' acts of hubris are inextricably bound with their pathology of greed (Greek: *koros, Ag.* 376–84).[35] The reflections of this ode, which implicate all three individuals in crimes that disturb the economic order – namely violation of the sacred in the name of profit, commodification and waste of human life with an eye to wealth, and insatiable greed – climax with the suggestion that the Erinyes may delay inflicting punishment, but eventually do punish such acts (*Ag.* 463–74):

> For the gods do not fail to take aim
> against those who have killed many, and in time
> the black Erinyes enfeeble him/who succeeds against justice,
> reversing his fortune and corroding his life.
> …
> I approve unresented prosperity;/I wish I may neither sack cities
> nor as captive/myself see my life under others. (tr. Sommerstein 2008, adapted)

The idea that the Erinyes will punish those who destroy human lives through their greed, excess and hubris is reflected upon further in the second choral ode of *Agamemnon*. At the ode's heart lies the contemplation of the woes that the unjust and greedy members of the elite bring upon the community.[36] Here, *Ag.* 750–76, which explicitly reflect on the unjust wealth of elite houses, point, at their climax, to the revenge of the *daimon* of the house. This divinity (*Ag.* 769–71), as Collard has rightly pointed out, is meant to be understood as the Erinys:[37]

> Ancient insolence is wont to breed/youthful insolence in evil men
> sooner or later, when the appointed day comes/for birth, rancour rising afresh,
> and a *daimon* unfightable invincible/unholy in boldness,
> a *daimon* of black Ruin/for a house, resembling its parents.
> Justice gleams in houses foul with smoke,/doing honour to the righteous man;
> but gold-bespangled mansions where hands are unclean
> she leaves with her eyes turned away …
>
> (*Ag.* 763–78)

The second choral ode thus powerfully prepares the way for the play's 'tapestry scene', where Agamemnon tramples on, and symbolically destroys, the intricate and highly expensive blood-red tapestries – the 'house's wealth'[38] – spread by Clytemnestra and the house's servants at his feet (*Ag.* 905–74). The dark red colour of this spectacular prop captures, as has long been acknowledged, the blood (and life) of both Iphigeneia and of countless youth who were sacrificed in the Trojan expedition by its leaders' greed and excess.[39] Agamemnon's trampling on the heavily symbolic red cloth conjures the earlier choral reflections about the elite houses having a destructive attitude borne out of excess. It is now that Iphigeneia's portrayal as the 'wealth of the house' makes full sense, corroborated by what the audience heard about the lives of the young soldiers, treated like commodities and wasted in a destructive war motivated by insatiability and greed. At the same time, the prop of the tapestry, which symbolically merges human life and wealth and presents them as subjected to wasteful destruction, ultimately serves as the Erinys' net that traps Agamemnon, leading him to his death for his crimes. For it is the Erinys, through the character of Clytemnestra, who lures Agamemnon into the wasteful act of destroying the tapestries and symbolically the human lives. The daimonic power, operating on behalf of the natural order which Agamemnon so hubristically abused, invites him to abuse not only the wealth of the house, but the wealth of natural productive powers more broadly:

> The sea is there – and who shall quench it? – nurturing the juices which yield much purple worth its weight in silver, wholly renewable, the dye of vestments. The house (*oikos*) has an abundance of these with the gods' help, my lord. This house does not know how to be poor.[40] (*Ag.* 958–65)

As, through the mouth of Clytemnestra, mouthpiece of the Erinys, we hear the confident assertion about the elite's entitlement to waste wealth, we see Agamemnon destroying the precious natural resource of purple dye and wasting the labour required by both nature and humans to produce this intricate and expensive tapestry. Through them, the image of the catastrophic waste of human blood – especially young blood, that of Iphigeneia and the soldiers expended at war – is evoked. The combination of words, dramaturgy and action in this striking evocation of the natural productive powers and their abuse, suggests that human life is a product of these powers; it is 'wealth', and it is abused like all wealth is.[41]

The choral ode that closes the 'tapestry scene' and subsumes its key ideas articulates the connection between wealth and human life much more clearly. The loss of material and natural wealth is here contrasted with the loss of human life. The earth can regenerate its natural wealth, even if this is destroyed, but once the wealth of lifeblood is shed on the earth, it is forever lost:

> Still, if caution casts forth/part of the goods in his possession
> from a sling of generous dimensions,/the whole house does not founder
> when crammed too full in surfeit,/nor does he wreck the ship:
> the gifts of Zeus are surely great, coming abundantly
> from furrows teeming year after year/to destroy the plague of hunger.
> But once the black blood of death/has fallen on the earth in front of a man,
> who by any incantation can summon it back again? (*Ag.* 1009–21)

What can be concluded from the passages above is that the *Oresteia* assumes and develops a strong connection between the Erinyes and wealth. The Erinyes are shown to be concerned with the abuse of resources, not only of material wealth, but also those coming from the natural productive powers, including the precious resource of human life. Beyond the choral odes, the 'tapestry scene' makes this strikingly clear, by foregrounding the heavily symbolic scene of abuse and destruction of the tapestries. Based on the above scenes, imagery and choral reflections, therefore, the Erinyes can be defined as chthonic not only because of their provenance from the earth, and certainly not only because of their connection with darkness, curses, monstrosity, terror etc., but because their sphere of activity is the guardianship of the 'wealth of the earth' as a precious resource and the punishment of its abusers.

Wealth, life and earth in Aeschylean drama: The chthonic in the *Persians* and the *Seven Against Thebes*

Many of the considerations with regard to wealth that emerge in the *Oresteia* have been developed and explored at length in earlier Aeschylean works, suggesting that this is a prevalent theme more widely in the Aeschylean corpus. In the *Persians* and the *Seven Against Thebes*, wealth amounts to an equally significant and complex concept, and its abuse is presented as nothing less than abuse of the cosmic order. This suggests something even more significant for our understanding of the *Oresteia*'s final scene: that the climax and closure of the *Oresteia* (Aeschylus' last work), where natural wealth is restored, respected and celebrated,[42] is also a climax and closure for a central preoccupation of the Aeschylean oeuvre.

In its clearest and earliest form, we find the connection between natural wealth, earth and cosmic order in the *Persians*, a play produced fourteen years before the

Reconsidering the Chthonic in Aeschylus' *Oresteia*

Oresteia (472 BCE). This play dramatizes the aftermath of the Persians' defeat by the Greeks in the sea battle of Salamis, an event whose significance, like the events in the *Oresteia*, is raised to one of cosmic proportions. From the very beginning, the narrated and dramatized action of the *Persians* constructs the Persian imperialist expedition as an epic struggle between humanity and natural environment. Persian imperialism, identified metaphorically with the hybristic human attempt to subjugate and dominate the environment (*Pers.* 66–80; 130–1; 181–96; 722; 746–8), is explicitly connected with insatiable pursuit of wealth (*Pers.* 754–8, cf. 842). The Persians' desire to possess unlimited land and wealth provokes a wrathful reaction by the natural order,[43] resulting in the destruction of earth's most precious wealth, human life. For, life itself, especially the youth of the Persians (*Pers.* 512),[44] mothered and nurtured by the earth,[45] is gradually, but emphatically, shown to be conceptualized as 'wealth', especially through imagery of agriculture and natural growth (*Pers.* 250–5; 821–6; cf. 925, 978). Having collapsed the element of life into other examples of natural production (cf. *Pers.* 618 'the children of the all-bearing earth'), and having conceptualized the earth as the ultimate economic source, the *Persians* suggests that youth is the greatest natural capital, the most valuable growth (*Pers.* 821–6). However, the striking imagery of the 'harvest of tears', whose enactment as dirge for the wasted youth pervades the last third of the play, poignantly envisages the catastrophic waste and destruction of this capital as caused by greed:

> Outrage has blossomed, and has produced a crop of ruin, from which it is reaping a harvest of tears ... let no one despise the fortune he possesses and, through lust for more, let his great prosperity go to waste (*Pers.* 821–6; cf. 842; tr. Sommerstein 2008, adapted)

The relationship between humankind and earth has gone horribly wrong, with humans having abused and wasted the earth's wealth and the earth having taken revenge with spectacular force. The 'harvest of tears' left at the end for man to reap has been brought about through the angry manifestation of all aspects of the natural environment. Land, sea and the other natural elements are shown to take revenge *together* on human hubris: Salamis' rocky landscape (*Pers.* 302–10) extinguishes the human wave of the Persian army (*Pers.* 88–90), the sea drowns the flower of Persian youth (*Pers.* 274–7), the fertile land deprives the retreating Persians of food and water (*Pers.* 488–91) and the sun melts the ice of the river condemning the remaining soldiers to a watery grave (*Pers.* 504–7).

Earth, human life, wealth and growth are once again prominent and interconnected ideas in Aeschylus' *Seven against Thebes*, the last (and only surviving) play of a tragic trilogy that dramatizes events leading to the fatal duel between the two brothers Eteocles and Polyneices over the possession of the rule of Thebes. As in the *Persians*, the concept of young life begotten by the earth is privileged from the outset (*Sept.* 16–20; cf. 70–5, 412–16, 474–9, 548, 584 etc.). The kinship between the notion of the 'earth' and that of 'motherland' allows the association of the city of Thebes and its earth to pervade the play (as e.g. in *Sept.* 70–5, 104–07, 167–9, 247, etc.).[46] Crucially, as in the *Persians*, possession and control of land is explicitly identified with possession and control of wealth (cf. *Sept.* 947–50, cf. 730–3, 815ff.).[47] Likewise, just as other natural growth is conceptualized as disturbed, so too is human life (*Sept.* 333–63), wasted and destroyed in an aggressive conflict motivated by the prospect of possession and gain (cf. *Sept.* 697 'gain comes before death, that comes after').[48] In this play, as before, Aeschylus depicts the conclusion of the struggle between humanity and earth in pathetic terms (*Sept.* 720–1004), with the two young brothers, Eteocles and Polyneices, last scions of the house of Labdacids, shown to have acquired only as much land (and wealth) as their bodies need for a grave:

> cruel-hearted Iron, allotting them land to dwell in, as much as is given to the dead to possess, with no share of the broad plains. (*Sept.* 730–3)

> They have received their allotted portion, the wretched pair, through suffering sent by Zeus; under their bodies there will be the limitless wealth of the earth. (*Sept.* 947–50; cf. 815–9, 901–14)

The brothers are shown to have reached 'limitless wealth' in the dark depths of the earth, but there it is of no use to them (*Sept.* 949). Thus this play also closes with the concept of the futility of the human pursuit of wealth at the expense of the earth, and especially at the expense of the earth's ultimate wealth, human life.

The chthonic, nature and the order of the cosmos: Towards a reinterpretation of the final scene of the *Oresteia*

The analysis above supports several valuable conclusions regarding the chthonic and the earth in Aeschylus. Earth appears to be a far more central, complex and multi-faceted concept in Aeschylean drama than previously thought, because it is deeply connected with a key concern in all Greek tragedy, human life and its loss – which in Aeschylean drama has a profoundly economic dimension. Earth in Aeschylus is imagined to be the source of human life, as it is the source of all natural and other forms of wealth. Earth is also the space of death and decay, to which everything returns. The *Libation Bearers* of the *Oresteia* captures this duality of the earth in the most succinct way:

> and Earth herself, who gives birth to all things, nurtures them,
> and then receives that fruit of her womb back into herself. (*Cho.* 127–9)

The depths of the earth, evoked as a persistent image in play after play, capture the idea of natural productivity. However, in spatial terms 'earth' – and by extension, the 'chthonic' – should not be understood as referring only to the land. This becomes clearest in the *Persians,* where, as we saw above, all aspects of the natural environment – from fresh to salt waters and from heavenly elements to fertile depths – conspire together to implement the 'harvest of tears', namely earth's revenge for the violation of its wealth. In the imagery of the *Persians,* the 'earth' that produces that sorrowful harvest is coextensive with the whole of nature and the cosmos.[49] Through the spatial and natural imagery of the play, Persian imperialism, which relentlessly seeks wealth and expansion, is portrayed as having violated the cosmic order. The cosmic order then turns against the expedition and annihilates it through the natural elements of sea, land, winds, stars and the depths of the earth. In this process, those same forces, in response to the Persians' greed and expansionism, poignantly annihilate their most precious wealth, human life.

The striking overlap between the 'cosmic' and the 'natural' that clearly emerges from the *Persians*[50] is entirely in line with philosophical thought in the archaic period, in which justice (Greek: *dike*) is inextricably connected to the natural order. In Hesiodic epic, on which Aeschylus draws, humans are urged to contribute to *dike* by living and working in balance with nature and by using wealth in accordance with the law. In sixth-century philosophy, nature is integral in ideas of justice, balance and order.[51] In Solon, the sea is said to be at its most just state (*dikaiotate*) when no wind disturbs it (fr. 12); elsewhere, the retribution that comes from unjust pursuits is portrayed as swelling until it lashes out like a wind (fr. 13.17–25); the earth is also imagined as violated due to the greed of the wealthiest classes who abuse the poor (fr. 36.1–15). In a fragment of Anaximander, the coming-to-be of all life is imagined as a natural process within a cosmos organized by law (A9, B1 DK). Perhaps most famously of all, in Heraclitus justice as cosmic order entails the sun not overstepping its measures (B94 DK). In these thinkers, the elements of nature that capture the cosmic order are presented as having animate status and agency,[52] and this is shared, as we saw, by the Aeschylean *Persians*.

This connection between the natural and the cosmic in Greek philosophers and in early Aeschylus is made even more complex in the *Oresteia*. Here, natural imagery not only reflects a reaction to the disturbance of the cosmos; the natural elements are often linked to, and sometimes identified with, divine agents who intervene to defend and restore the cosmic order: namely the Erinyes. The terrible weather conditions which plague the imperialist and destructive expedition to Troy from the start (*Ag.* 185–218, 528, 555–69), the animal world that reacts with hostility corroborating the weather's reactions (*Ag.* 111–21, 134–7, 717–36), the storm that annihilates the expedition's ships on their return (*Ag.* 505, 634–77), the spring of blood that ominously irrigates the thirsty earth when Agamemnon is murdered (*Ag.* 1389–92), and the storm and winds that wreck the house of the Atreid family (*Ag.* 1533–6, *Cho.* 1065–74), are examples of imagery depicting the disturbance of the natural order in the *Oresteia*, as many have noted.[53] However, what makes this imagery more striking, especially in contrast to the natural imagery in the *Persians*, is the discernible link with the Erinyes behind the agency and punitive effects of these natural phenomena and elements. In other words, in the *Oresteia* natural imagery and the operation of the Erinyes are linked in a complex, and increasingly self-reinforcing manner, as they manifest their agency as guardians of the cosmic order and avengers of the abuse of the earth's wealth.

The most consistent manifestation of this is through storms or winds that, literally or metaphorically, cause wreck, destruction or madness. Although initially the link between winds, storms and the Erinyes appears mostly in a cryptic form, it becomes more explicit as the trilogy unfolds, through a complex nexus of cross-references. A clear example of this is the storm that the Erinyes claim to stir in order to punish human greed:

> But he, I say, who audaciously acts contrary to this, lawlessly carrying a heavy cargo heaped up in confusion, against Justice, will in the end be forced to lower sail, when trouble takes hold of him, by the shattering of his yard-arm. In the middle of the eddies, unable to fight his way out, he calls, but they pay no heed; the *daimon* laughs at the headstrong man, seeing him powerless, the one who boasted it could never happen, in helpless distress, as he fails to surmount the crest of the wave: he has wrecked the ship of his former lifelong prosperity on the reef of Justice and perishes unwept, unseen. (*Eum.* 553–65)

Throughout the *Oresteia* we find mention of storms and winds that battle human endeavour, and they have a consistent connection with the Erinyes: for example, the storm that wrecks the returning fleet of the Argives and is understood to be a victory song to the Erinyes (*Ag.* 645–67); the winds that bring Helen-Erinys to Troy (*Ag.* 688–98, 737–49), the spectres that whirl Orestes about like a storm (*Cho.* 1051–2), and who are literally the chorus of servant women-Erinyes;[54] the scorching breaths that Clytemnestra orders the Erinyes to direct at escaping Orestes (*Eum.* 137–39, cf. *Eum.* 328–33, 840), and many others.[55] *Eum.* 553–65, therefore, where the Erinyes appear unambiguously as controlling and manifesting through winds and storms, is no doubt the climax of a representation that emerges earlier, albeit in more cryptic forms.[56]

Beyond winds and storms, other examples of the Erinyes operating through disturbed natural imagery include the agency of animals. In the parodos of *Agamemnon*, for example, the Erinys is ultimately identified with the birds of prey Agamemnon and Menelaus who devour the pregnant hare (*Ag.* 48–62, 111–21). In the second choral ode of the *Agamemnon*, the lion cub pet that massacres its family eventually identifies with the Erinys of the house (*Ag.* 717–49).[57] In the agricultural metaphor of blood irrigating the thirsty soil at the murder of Agamemnon, the agency of the Erinys of the house emerges clearly behind the voice of Clytemnestra (*Ag.* 1389–92).[58]

From the passages above, we can deduce that the construction of the Erinyes as operating within and through the natural world suggests that chthonic and natural are inextricably connected with one another, and, more crucially, have a key role in the maintenance of the cosmic order.[59] Even more importantly, it can also be concluded that for these deities, the axis of the cosmic order and its disturbance is use and abuse of the earth's wealth. In the Aeschylean worldview, cosmic balance, through the operation of the Erinyes, is threatened by abuse and misuse of precious wealth, all of which ultimately belongs to the earth. This balance, which is also identified with *dike,* justice, in Aeschylean tragedy must be guarded, and violation against it punished. This is precisely the role that the Erinyes perform until the last scene of the trilogy: they are guardians of the wealth of nature, earth and the cosmos. However, in the process of defending the natural order, redressing and punishing its violations, the Erinyes generate more violations, perpetuating the endless cycle of violence, destruction and revenge. This is the problematic state of affairs in the trilogy that needs to be addressed; and this comes in the very last scene, to which I finally turn.

The focus on the Erinyes as agents of natural wealth and life in the final, climactic scene shows that a central concern of the entire trilogy is balance within the cosmos through respect for the earth's wealth – the wealth which, throughout the *Oresteia*, as well as in earlier Aeschylean dramas, has been misused and abused. The Erinyes are thus accepted in the new order not only for the sake of balancing the upper with the lower realms (although this is undoubtedly important). They are shown in their full glory, as chthonic deities of wealth whose realm stretches well beyond the ground, and includes nature and, by extension, the cosmic order. This is what Athena recognizes when she accepts the Erinyes as incoming residents in the Athenian land:

> Athena (to the Erinyes): Such [blessings] as are appropriate to an honourable victory, coming moreover both from the land, and from the waters of the sea, and from the heavens; and for the fruitfulness of the citizens' land and livestock to thrive in abundance, and not to fail with the passage of time; and for the preservation of the human seed. But may you give greater fertility to those who are pious; for like a shepherd of plants, I cherish the race to which these righteous men belong. Such things are for you to grant. (*Eum.* 903–13)

With Troy, Argos and Delphi, where the deities' wrath and destructiveness were fuelled, left behind, the Erinyes are envisaged as having reached their permanent seat in the land of Athens, the only space where their energies can literally and metaphorically flourish. In the blessed natural conditions of Athens and in reward for the justice of its citizens, the deities will act as protectors and guarantors of the cosmic order. By ensuring the protection and flourishing of life, natural production and wealth, the threat to human life and to all of the earth's wealth – largely caused by greed, insatiability and destructiveness – will be greatly diminished. If life is threatened, then both the deities and their civic analogy, the newly instituted Areopagus, which they oversee, will ensure the restoration of order. Ultimately, in the final climactic scene of the *Oresteia*, the land and polis of Athens is envisaged like a microcosm, from which the chthonic Erinyes, in symbiosis with the patron-goddess Athena, will ensure the order of the larger cosmos. In this space, violations of earth and its wealth, played out relentlessly in *Agamemnon* and *Libation Bearers*, will ultimately stop.[60]

The chthonic in the *Oresteia*, therefore, and probably in the whole of Aeschylean drama, should be regarded as a far more complex notion than we thought until now. A reading of the Aeschylean plays with a focus on earth and its wealth suggests that Aeschylean drama attests to the existence of what we would today identify as ecotheology. Put differently, the Aeschylean representation of divine agents that avenge abuse, waste and destruction of natural wealth and promote growth when order is maintained shows a conscience which today we would characterize as ecological. In the Aeschylean context, far from being irrelevant to modern ecological concerns, the chthonic embodies the sentiment that it is the balance between humans, earth and its wealth that guarantees the balance of the cosmos. Viewed in this way, the Aeschylean conceptualization of the Erinyes, the chthonic, the earth and their operation becomes strikingly relevant to our modern capitalist, destructive and wasteful societies.

CHAPTER 9
THE ANGUISH OF THE EARTH: ECOLOGY AND WARFARE IN THE FIRST WORLD WAR AND THE BIBLE
Hilary Marlow

Introduction

Throughout human history, warfare and armed conflict have inflicted substantial damage on human communities and settlements, in addition to the direct losses sustained by fighting armies. In the twentieth century alone, it is estimated that between 135 and 148 million human lives were lost in conflicts worldwide, plus of course, the countless other people seriously wounded or disabled (Leitenberg 2006).[1] Fighting, invasion and siege warfare also seriously impact people's livelihoods, especially in non-industrial societies, as agricultural produce is destroyed or commandeered to feed troops, and other natural resources such as timber and coal are taken to supply the seemingly endless demands of the military machine. Small wonder then that wars increase the cycle of poverty and deprivation especially among those on the margins, leading to further deaths, and, in contemporary societies, an increased reliance on foreign aid (Goodhand 2003: 631–2).

In addition to the devastating effect that war has on human communities, the natural world itself suffers drastic damage in armed conflicts.[2] Yet the destruction of ecosystems and wildlife habitats in war zones, and the attendant casualties among animal populations, often receive little attention from commentators. In modern warfare the environmental effects of war may be felt directly (for example damage from shelling, scorched earth policy or chemical spraying) or indirectly (for example long-term pollution of land and water, or deforestation). But such destruction is not just a problem associated with modern warfare. Ancient Greek authors record explicitly and implicitly damaging practices such as burning scrub cover (e.g. Thucydides, *History of the Peloponnesian War* 4.29–30, 38; Herodotus, *The Histories* 6.78–80) and damage to forests (e.g. Diodorus Siculus, *Library of History* 14.42.4; Athenaeus, *The Deipnosophists* 5.206f, 208e–f), and the destruction of trees during siege warfare is expressly prohibited in the Hebrew Bible (Deut. 20.19–20, see also 22.6–7). The extensive deforestation of the Mediterranean basin during the Graeco-Roman period, which permanently altered

---

The author is grateful for constructive comments on earlier versions of this chapter presented at the Faraday Institute Research Seminar in 2016, and the Book and the Sword Workshop in 2017

the landscape, was in part at least in order to provide fuel and materials for various war efforts (Hughes 2013; Hughes and Thirgood 1982).

Sometimes the opposite can be true: conservation biologists have noted that the absence of human populations as a result of warfare and the creation of 'no man's land' between military lines can be beneficial for wildlife and ecosystems (Martin and Szuter 1999). Some studies of the effects of warfare on the physical environment have demonstrated the resilience of this landscape and its remarkable ability to recover from major anthropogenic disruption (Hupy and Schaetzl 2008). However, even this apparently positive side effect may present problems, and modern methods of warfare increasingly cause widespread destruction and long-term damage.

One salutary example of this from recent history is the enduring effect that the First World War had on the physical landscapes of France and Belgium. The heavily forested areas surrounding the battlefields of the Northern Front, stretching from the North Sea to Switzerland, were extensively destroyed to provide the vast quantities of timber needed to support the war effort. Between 70 per cent and 80 per cent of the mature oak and beech forests of Northern France were razed to the ground, to supply fuel and construction materials (Dubois 1994: 166). One estimate suggests that the extent of soil erosion in lower forest slopes was the equivalent of between 10,000 and 40,000 years worth of natural erosion (Puyo 2004: 578).[3]

Perhaps surprisingly, the environmental impact of this 100-year-old conflict continues to this day, in the form of tons of unspent munitions and acres of contaminated soil in France and Belgium that present a very present danger to life. To give one example, in 1916 the area around the French city of Verdun was subjected to ten months of intense artillery bombardment, in which over 300,000 men died. Military experts estimate that, of the 60 million shells fired, one in five ammunition rounds failed to detonate, leaving a treacherous legacy. As a consequence, after the war ended 16 million acres of Northern France were cordoned off from public access because of the dangers from an estimated total of 12 million unexploded shells. Astonishingly, these areas remain out of bounds to this day, while teams of *démineurs* or bomb disposal experts continue to clear the ordinance, but it is a hazardous occupation, sometimes resulting in death or serious injury.[4]

In addition, at the end of the First World War thousands of tons of unused munitions were dumped in numerous watercourses and lakes throughout Northern France. Some of these stocks were destroyed in the immediate post-war years resulting in severe soil contamination, others remain to be cleared (Bausinger et al. 2007). Although significant levels of toxic metals such as arsenic, lead, phosphorus, cadmium and mercury have been identified by researchers (Bausinger et al. 2007; Van Meirvenne et al. 2008), few studies have been conducted on the possible contamination of local water sources by decaying weaponry. Such biological analysis and statistical data demonstrate the very many ways in which the effects of the First World War, which so profoundly altered the historical and political map of Europe, also damaged irrevocably the ecological contours of the battlefields of France and Belgium, producing effects that continue to be felt today in the natural environment.

Ecology and warfare in poetic texts

Interesting as such evidence of the devastating effect of the First World War on landscapes and communities may be, this is not what captures the public imagination. The trauma and devastation of this war are known to us primarily through the evocative and shocking outpouring of poetic verse that it generated. Jon Stallworthy cites Wordsworth's description of poetry as 'the spontaneous overflow of powerful feelings' before adding 'there can be no area of human experience that has generated a wider range of powerful feelings than war: hope and fear; exhilaration and hatred' (Stallworthy 2014: xxii, see also Eberhart and Rodman 1945). It should come as no surprise then that those on the front line poured out their emotions in poetic form. As Stuart Lee puts it, 'The First World War provides one of the seminal moments of the twentieth century in which literate soldiers, plunged into inhuman conditions, reacted to their surroundings in poems' (Lee 1996).

However the First World War poets are by no means the first authors to paint graphic pictures of the effects of war, nor to draw attention to its impact on individuals and communities, as well as the physical landscape. From the great epic poems of ancient Greece and Rome through the Norse sagas to the poetry of the American Civil War, the violence and trauma as well as the victories of war have been captured in verse. This is nowhere more evident than in the poetry of the Hebrew Bible, where the effect of warfare and enemy invasion on the land and its inhabitants is a significant feature of oracles in the prophetic books, as the following examples demonstrate:

> The destroyer shall come upon every town,
>   and no town shall escape;
> the valley shall perish,
>   and the plain be destroyed,
> as the LORD has spoken. (Jer. 48.8)[5]

> Let the sword fall twice, thrice; it is a sword for killing.
> A sword for great slaughter – it surrounds them;
> Therefore hearts melt and many stumble.
> At all their gates I have set the point of the sword.
> Ah! It is made for flashing, it is poised for slaughter.
> Attack to the right! Engage to the left! (Ezek. 21:14–16)

In the remainder of this chapter I will examine references to the environmental damage resulting from war in the prophetic texts of the Hebrew Bible and bring them into conversation with the poetry of the First World War.[6] At first glance it might seem an arbitrary choice. Why bring these two significant literary collections together in this way? And why focus on damage to the physical environment? In the first place, as a biblical scholar I am especially interested in the multiple ways that the Bible is reused in later literary works, either explicitly or implicitly, through the reframing of themes and motifs from biblical texts for later contexts. In the highly religious society of late Victorian and early Edwardian England, children were schooled in the Bible from an early age, both in the home and at church. So one interesting consideration is whether and to what extent the First World War poets drew on biblical ideas, whether consciously or unconsciously. However, many of the educated soldier poets of the era would also have been well schooled in the Classics. So it is highly likely that the poetry they produced would be imbued with classical allusions as well. The study will draw attention to such cultural appropriation and transmission, both biblical and classical, before discussing how much of this is intentional.

Second, my interest in contemporary ecological hermeneutics makes me attentive to the language used to depict the earth and its fate at the hands of human agents, which is a feature of both the Hebrew Bible and the First World War poets. The lasting legacy of ecological damage inflicted on the battlefields of the First World War is a reminder of the devastation that human beings cause to the planet. Although in the present era we are more aware than ever of the environmental consequences of our actions, the biblical prophets writing two and a half millennia ago were alert to human-induced damage to the earth. Although there are some similarities in the ways in which the effect of war on the physical world is depicted in the two collections of war poems, there are also significant differences. In what follows we will first highlight a number of key features and recurring themes of biblical 'war poems' that portray the impact of war on the land.[7] We will then consider whether such features also characterize the poetry of the First World War and the reasons for the similarities and differences between the two sets of material. Finally we will ask what, if anything, these two significant literary collections can contribute to current ecological thinking.

War in the Hebrew Bible

Ancient Israel has been described as 'one of the main military thoroughfares as far back as written annals record' (Herzog and Gichon 1978: 12). Its unique location sandwiched between Mesopotamia to the east and the Mediterranean Sea to the west made it invaluable in the ancient world as a trading route connecting not only east and west, but also providing the only land bridge linking Africa in the south with Europe and Asia. It enjoyed relative fertility, compared with the desert lands to the south and east, and thus was an attractive proposition for would-be settlers as well as providing rich pickings for raiders from adjoining nations. It is unsurprising then, that political 'superpowers' in succeeding periods were willing to conduct extensive military campaigns to secure this narrow stretch of land.

The first such recorded campaign is the Egyptian incursions into Canaan by Pharaoh Pepi I, dated to the Intermediate Bronze Age (*c.* 2400 BCE) and recorded on the tomb inscription of his general, Uni (de Miroschedji 2012; Herzog and Gichon 1978). With varying degrees of potency, the Egyptians remained the dominant military force in the region until the collapse of Late Bronze Age Canaan. By the end of the eighth century the southern Levant was threatened by a new imperial power, the Neo-Assyrian empire, which '[by] maintaining a consistent presence in the region and occasionally using force … forged the "global" empire of its day' (Golden 2004: 169). By that stage, according to the biblical record, the Israelite settlers in Canaan, the 'promised land', were divided into two nations, the Northern Kingdom of Israel and the Southern Kingdom of Judah. Among the Assyrian conquests was the Northern Kingdom of Israel, defeated and exiled by the Assyrian king Tiglath Pileser III in about 720 BCE (see 2 Kings 17). After some seventy years, the Assyrians themselves fell victim to the rise of the Neo-Babylonian empire from further east, which was responsible for the conquest, and subsequent exile, of the Southern Kingdom of Judah at the end of the sixth century (see 2 Kings 24–25).[8]

Biblical war poetry

It is against this historical backdrop of conquest and reconquest that a significant number of the poetic and prophetic texts in the Hebrew Bible are set. Many of the oracles in the written prophets are an impassioned response to political threat or military invasion and defeat, from the eighth century BCE onwards. Although the texts often furnish little in terms of factual details or historical understanding of the events themselves, many paint graphic and detailed pictures of the experiences of war and invasion on communities, and as we shall see, on the physical earth. There is little way of knowing whether the authors were themselves witnesses of the scenes that they portray or were drawing on collective memories and accounts handed down within their communities. However, it is unlikely, given the passion with which they speak and the details they convey, that they are conjuring up imaginary scenes of violence and destruction.

These prophetic texts typically bear the characteristics of Hebrew poetry, using parallelism, metre and other mechanisms to create literary art that will engage and move the hearers or readers (Henderson 2007: 118–19). The texts come in various literary forms, ranging from laments to pronouncements of coming divine judgement – either addressed to the people of Israel and Judah, or to those who have oppressed them. The authors use graphic imagery and hyperbolic language to portray scenes of devastation and pain. Examples include the prophet Nahum's depiction of the destruction of Nineveh (Nah. 2.1–12); Jeremiah's focus on the coming invasion from the north (Jer. 1.13–16, 4.5–8, 6.22–26) and Isaiah's oracles against Babylon (Isa. 12.1–22), and Moab (Isa. 15.1–9). A surprising number of such texts portray the environmental impacts of warfare as well as its social and political dimensions, either by describing the physical reality of war-damaged landscapes or by means of extended and evocative figurative language, or both.

Unlike poems of the First World War, which often draw on the experiences of soldiers in the trenches and life at the front line, many biblical war poems are written from the perspective of ordinary people enduring the trauma of invasion or siege. The emotional tenor set by these poems aptly conveys the terror and fear that war generates for the human inhabitants of the land. It is important to note too that many of these texts segue seamlessly between war as conflict between human armies, and war as the judgment of Israel's God YHWH on other nations, or even on his own people.[9] This reflects ancient views on the overlap between human affairs and the interests of the gods, and the understanding throughout the ancient Near East that deities governed a particular geographical and national space. The causal relationship between human and divine spheres, whereby human wickedness results in punishment by the gods, characterizes the thought of the Hebrew Bible and is highlighted by the prevalence of language of YHWH as warrior (e.g. Isa. 42.13; Joel 2.11). In many of the biblical prophetic war texts, the human battle on earth is an enactment of what has been determined by YHWH and his cosmic forces, and is often heralded as the coming 'Day of YHWH' (e.g. Isa. 13.9; Joel 2.1). Given the Hebrew Bible's emphasis on the land as a gift from YHWH, it is unsurprising then that many of the Day of YHWH texts are concerned with the implications of warfare for the land and its fertility (e.g. Joel 1.15–20; Zeph. 1.14–18, cf. v.3).

This poses interesting questions concerning the ethics of the texts and the actions of YHWH, which potentially have implications for an ecological reading of the texts. Does Israel's God deliberately damage the land that is his creation and his gift to the Israelites? Does this legitimize human exploitation and destruction of the earth? This is the charge laid at the biblical text by the work of the Earth Bible Project, founded in 2000 by Norman Habel and Peter Trudinger of Flinders University, Adelaide. Habel suggests that the earth suffers injustice at the hands of both humans and God, and argues for a resistant counter-reading of the dominant patriarchal anthropocentric orientation of the biblical texts (Habel 2000: 40; see also Habel and Trudinger 2008: 5). There is not space here to discuss this in detail but I have argued elsewhere that the Earth Bible approach is highly subjective and anachronistic, and may invite rejection of the biblical text as a source of ecological reflection (Marlow 2009: 90–4). Habel's work takes no account of the biblical world view in which actions have consequences, and where ancient Israel is promised the blessings of peace and a fertile land if they obey YHWH but curses and a ravaged land if they disobey (see Deuteronomy 28).

Returning to the prophetic war poems, we will now examine a number of key motifs and metaphors that highlight the effect that war has on the physical earth, whether as a result of human or divine action.

Earth mourning

Biblical war poems convey the anguish of war in emotionally raw terms, often using the form of lament (e.g. Jer. 4.19–20, 6.26; see also Lamentations, esp. 1.4, 2.5, 5.15).[10] Despite the fact that warfare is primarily a male activity, the analogy of the pain of childbirth is frequently used (e.g. Isa. 13.8; Jer. 6.24). This may be an acknowledgement by the male authors of the severity of labour pains endured by their women, but may also relate to the female personification of Lady Jerusalem in Hebrew poetry, for example in Isa. 3.26, where she is ravaged by the act of war and left mourning and desolate (see also Lam. 1.4).

In these examples of lament, the grief is that endured by the people, but on occasions the earth itself is identified as a grieving victim – of human sinfulness, societal violence or warfare. In nine distinct prophetic passages in the Hebrew Bible, the act of mourning or lament is attributed to the earth, using the Hebrew verb 'ābal in conjunction with 'ereṣ 'earth'/'land' or an equivalent (Hayes 2002).[11] The verb 'ābal is normally translated as 'to mourn' when used of human subjects, but in some of the texts in which it is paired with 'ereṣ it carries the association of 'to dry up' by virtue of the parallel verb yābēš, thus signifying the destructive effect of prolonged drought (e.g. Jer. 12.4, 23.10; Amos 1.2, see Hayes 2002: 12–18.). This recurring motif in texts ranging from the pre-exilic to postexilic eras is a root metaphor that links the state of the earth to that of the human community, whether Israel/Judah or the wider inhabited world, in which, suggests Hayes, 'earth assumes a persona, responding to human distress or transgression (or to both)' (Hayes 2002: 4). The modern science of ecology is based upon the complex web of links between all living things, including humans and their habitats. These ancient texts also assume a level of interconnection between people, other species and the earth, but introduce a moral component that highlights the culpability of human beings in the devastation of the physical landscape.

This moral ambiguity is particularly evident in some passages identified by Hayes, in which the mourning of the earth is linked, either explicitly or implicitly, to the impact of warfare, whether human activity or the coming of YHWH as warrior. For example the extended metaphor of Joel 1–2 likens the devastation of an invading army (2.20) to a plague of locusts destroying everything in its wake (1.4) and causing the ground to mourn (1.10). Here the interplay between the human army, divine activity in the form of the Day of YHWH and the locust plague is deliberately left ambiguous (Simkins 1991). Similarly, in Hosea 4 human actions have a serious impact on the wellbeing of the land. The Israelites' neglect of YHWH (4.1) causes society to disintegrate (4.2) and the land to mourn and wild animals to perish (4.3). The emphatic *al-kēn* 'for this reason' at the start of verse 3 emphasizes a clear causal relationship between the indictment of the people and the devastation of the land, although it is unclear whether this is a result of warfare, severe drought, or some other environmental disaster, actual or imagined. Another example occurs in Isa. 24.1–13 in which YHWH is clearly the agent; it is his judgment in response to human sinfulness that results in a mourning and polluted land (vv. 1–3). Although the parallel between *ābal* 'to mourn' and *nābēl* to 'droop' or 'wilt' in verse 4 might suggest that a severe drought is in view, the depiction of the ruined and battered city in verse 12 speaks of military defeat.

In the book of Jeremiah, two examples of the earth mourning are linked with warfare, namely Jer. 4.23–28 and 12.7–13. Although the reason for the destruction is the physical invasion in both cases (4.5–8, 19–22, 12.12), a clear causal link is made with divine anger and judgement in chapter 4 (vv. 18, 28) while in chapter 12, it is the failures of Israel's political and religious leaders, the 'shepherds', that is to blame (12.10). For these Israelite prophets then, ecological devastation has its roots in hostile human behaviour either towards the land itself in warfare or towards their God.

Personification of nature

The notion of the earth mourning is one example of many in the Hebrew Bible in which human actions and emotions are apparently attributed to non-human natural elements. The reasons behind such personification of nature are debated, and, as I have argued elsewhere, the phenomenon should not be dismissed as 'merely' a literary device (Marlow 2013). Many of the poetic examples of personification are a positive summons to all living things to worship YHWH, for example the exhortation in Ps. 150.1 'Let everything that has breath praise the LORD!' Even inanimate natural elements are included in the call to praise their creator (Psalm 148, Ps. 98.7–8). However, there is also a darker side to the way that nature is personified in biblical texts. In language evoking a law court, heaven and earth function as witnesses to the broken covenant between YHWH and Israel (Deut. 30.19, 31.28, see also Isa. 1.2) and mountains and hills are invoked in YHWH's trial of his people (Mic. 6.1–2). Elsewhere the earth acts at YHWH's command to be the agent of his judgement against human populations, for example through earthquake (Amos 8.8) or flood waters (Amos 9.6).

In several biblical war texts, the earth is personified in other ways besides the language of mourning – either as the victim of war or as a salutary reminder of its consequences. So, in the book of Joel the earth 'quakes' in the face of the destroying army, whether human or locust (Joel 2.10). In Jeremiah 4, after the land has been utterly devastated (v. 20), the mountains and hills shake (v. 24) – the only movement in a desolate landscape. In the oracle against Babylon later in the book, the earth 'trembles and writhes' as the destruction of the enemy looms (Jer. 50.29). Similarly, in Isaiah the physical earth is shaken (Isa. 13.13), or staggers and falls like a drunkard (24.18–20), in anticipation of the coming judgement of YHWH. Micah also graphically portrays the effect of judgement on the physical earth: 'Then the mountains will melt under him and the valleys will burst open, like wax near fire' (Mic. 1.4).[12]

Cosmic disorder and de-creation

Some texts use images of cosmic disorder to depict the horror of battle. In Isa. 13.10, the sun, moon and stars cease to give light in a reversal of the Genesis creation ordinance that established the certainty of day and night (Gen. 1.16–18, 8:22). Later in Isaiah, in a text depicting the coming judgement of YHWH against all nations (Isa. 34.1–4), the heavenly bodies will rot away, and 'the skies roll up like a scroll' (34.4).[13] The tone set by these texts (considered by most scholars to be post-exilic) is a foretaste of the apocalyptic rhetoric depicting the end of the natural world that will characterize later Jewish and Christian writings (e.g. 2 Esd. 5.4–7; Mt. 24.29; Lk. 21.26).

A number of texts depict what scholars term the 'unmaking' of creation as a result of war (DeRoche 1981; Holladay 1961). For example, in Zephaniah, the order in which living creatures are created in the Genesis 1 account – fish, birds, animals, humans – is reversed (Zeph. 1.2–3) to demonstrate the catastrophic nature of the coming Day of YHWH (DeRoche 1980). Jeremiah 4.23–28 demonstrates this phenomenon in more detail, and specifically in the context of human warfare.. The section forms part of a series of dialogues between YHWH and the prophet Jeremiah that focus on the impact of enemy invasion (the 'evil from the north' of 4.6), and it 'imagines the world on the brink of extinction' (Stulman 2005: 64). A number of scholars regard it as an insertion into earlier historical experiences by later communities – an apocalyptic vision that depicts 'the mythology of chaos' (Carroll 1992: 168; see Childs 1959: 197).

In these verses the earth reverts to a pre-creation state of chaos and darkness (Jer. 4.23), in which both humans and birds have fled (v. 25). The use of language found in both Genesis creation accounts, in particular the phrase *tohû vābohû* ('waste and void' v. 23) that only otherwise occurs in Gen 1.1, suggests that the author is intentionally reversing creation traditions (Craigie et al. 1991: 81; Hayes 2002: 84). The impact of the horror is emphasized by the fourfold repetition of 'I looked' (*rā'îtî*, vv. 23–26) giving rhetorical force to a 'haunting first-person visionary account' (Biddle 1996: 22). The 'staccato rhythm of shock' created by this emphasis on the prophet's act of looking paints him as 'a helpless witness of disaster, a horrified onlooker, powerless and traumatised before the unfolding cataclysm as the earth turns into a lunar surface and a bombed-out landscape' (O'Connor 2011: 52).[14]

Although the prophets of the Old Testament were blissfully unaware of the kind of environmental issues that threaten the planet today, they nevertheless give voice to a powerful and enduring moral principle, that actions have consequences, often serious ones, and that these have a far-ranging effect on the well-being of others, human and non-human alike. In that respect then, they have enormous contemporary relevance.

First World War poetry

What then of the First World War poets? They are much closer in time to our own age, yet without the same level of environmental awareness that characterizes our society. In this section of the chapter we examine the work of these poets from the perspective of the connections made between human violence and bloodshed, and the earth.

The poems of the First World War cover the full spectrum from victorious, almost spiritual, triumphalism in the face of danger to melancholy musings or embittered outbursts on the futility and alienation of war.[15] Although many of these evocative descriptions deal with human suffering and paint graphic pictures of troops and dying men, of trenches and guns and other paraphernalia of war, some include poignant, if passing, references to the ecological damage wrought on the earth itself. Indeed, the title of this chapter, 'The Anguish of the Earth' is itself taken from one of Siegfried Sassoon's early poems, *Absolution* (Copp 2001: 78).[16] Being attentive to these allusions as well as to the physical conditions of the land in Northern France that still bears the scars of war speaks powerfully to our present situation where human damage of the planet, including the collateral damage from warfare, is reaching unprecedented levels.

The earth in First World War poems

As in the Old Testament, nature features in a significant number of poems and in a variety of ways, from brief allusions and snatched fragmentary snapshots to more intense similes and descriptions. Paul Fussell observes 'since war takes place outside and always within nature, its symbolic status is that of the ultimate anti-pastoral' (Fussell 2000: 231). The long pastoral tradition of English writing surfacing in a number of First World War poems serves, suggests Fussell, both as a measure against which to gauge the horrors of war and also as an imaginative protection against them (Fussell 2000: 235).

This is demonstrated in a notable feature of many of the poems, namely the contrast drawn between the idyll of green and pleasant England, and the derelict and barren landscape that the soldiers encounter on arrival at the Front:

> They march from safety, and the bird-sung joy
> Of grass-green thickets, to the land where all
> Is ruin, and nothing blossoms but the sky
> That hastens over them where they endure
> Sad, smoking, flat horizons, reeking woods,
> And foundered trench-lines volleying doom for doom.
>
> Sassoon, 'Prelude: The Troops' (Sassoon 1984: 67)

The horror of war extends to the view of the physical landscape that greets the soldiers arriving at the front, and is in direct contrast to 'England's green and pleasant land' that they have left (Blake 1993 [1810]).

However, relatively few poems make any extended reference to the decimation of nature and it is unclear to what extent those that do so are consciously cataloguing the damage caused by war. Rather nature images are deliberately woven into their poems to graphically enhance their picture of human misery. An example of this is Robert Nichols' account of going over the top, which paints a terrifying picture of the horror of trench warfare, and includes allusions to the effect of it on the landscape.

> Black earth, fountains of earth rise, leaping,
> Spouting like shocks of meeting waves.
> Death's fountains are playing,
> Shells like shrieking birds rush over;
> Crash and din rises higher …
>
> Go on. Go.
> Deafness. Numbness. The loudening tornado.
> Bullets. Mud. Stumbling and skating.
> My voice's strangled shout:
> *'Steady pace, boys!'*
>
> Nichols, 'The Assault' (Nichols 1917: 41)

Such vivid imagery highlights the author's sensory awareness – sight, hearing, touch, something that, as we have seen, is also important in Jeremiah 4. The earth is violently thrown into the air as wave after wave of shells hit, and the soil will soon be contaminated by the blood of the dead. Here environmental damage to the earth is clearly a by-product of trench warfare and human death.

Nevertheless, a number of the poems do articulate more clearly the effect of war on the land in its own right. Is it just coincidence that many of these use forms and themes already identified in biblical war poems, namely the personification of nature, the language of earth mourning and the notion of cosmic distress? We will discuss each of these in turn now, and then turn to a motif that is largely absent from the Hebrew Bible but more prevalent in classical texts, namely that of mother earth.

Personification of nature

The personification of earth features in several descriptions of the bleak and damaged landscape, such as Stuart Bellhouse's evocative and depressing picture of a new day:

> In Flanders where the soldiers lie,
> The mist-hung world is cheerless grey,
> And willows watch with leaden eye
> Penurious dawn greet haggard Day.
>
> Bellhouse, 'Two Mornings' (Copp 2001: 111)

Aimee Byng Scott's short two-part poem depicts first an idyllic landscape of bright-red poppies, ripening corn and the distant song of a skylark, before describing the death of nature as the bombardment starts up:

>   A shuddering night;
> Flames, not of poppies, cleave the quivering air,
> The corn is razed, the twisting trees are dead;
>   War in his might
> Has passed; Nature lies prostrate there
>   Stunned by his tread.
>
> Byng Scott, 'July 1st, 1916' (Reilly 1981: 97)

Some poems paint an evocative picture of damage and desolation, such as Vivian T. Pemberton's vista of damaged trees that 'bow like aged men' among the serpentine trenches:

> On the right the long white trenches,
> Waterlogged and full of stenches
> Crawl like snakes across the Lens-La Bassee road,
> Where the trees all bent and battered,

Seared with scars where shrapnel spattered,
Bow like aged men beneath a heavy load.

Pemberton, 'To Wingles Tower' (Copp 2001: 112)

Others are just fleeting references to the land that form a backdrop to the author's focus on the human misery and suffering engendered by war, for example Kenneth Saunders' depiction of the patience (and perhaps helplessness) of the polluted landscape:

The patient earth
Lay fouled with blood and riven with ruinous shard

Saunders, 'The Three Crosses' (Copp 2001: 89)

In a more specific metaphor, Mary Borden, who ran field hospitals in France during the war and was one of a number of female war poets, likens the landscape to one of her hospital patients:

Picardy is shaking with a fever,
Picardy's hills are wounded and broken,
Picardy's fields are scarred as with smallpox.

Borden, 'Where is Jehovah' (Borden 2015: 27)

The imagery that such figurative language conjures up serves to highlight in stark and evocative terms the damage to the physical landscape. In the light of the comparison between biblical and First World War material in this essay, it is interesting that the opening lines of Borden's poem make direct reference to biblical themes, calling on 'Jehovah, the God of Israel with his Ark and Tabernacle'. She likens the battle zone to the hubbub of the Israelite army camped in the wilderness: 'Here is a land that was silent and desolate,/suddenly covered with noise and confusion' and the noise of the guns to the thunder of Jehovah from the mountaintop:

This is His Hour, but Jehovah has missed it.
This is not His thunder nor His lightning.
These are not His people.
These are the armies of France and of England.

Borden, 'Where is Jehovah' (Borden 2015: 27)

Earth mourning

One aspect of First World War trench warfare that features in many poems is the cold, seemingly endless rain that turned the battlefields into swamps and made life for the soldiers such a misery. This is of course a contrast to the motif of 'mourning through drought' that many of the Hebrew Bible texts articulate. However, drought is a feature in several poems where both humans and the earth are parched and cry out for the relief that water brings:

> The sun beat down like force of hell and the earth was brown and dry.
> Tom shouted out for water while the earth called out for rain.
>
> Digby B. Hasler, 'Tom' (Copp 2001: 135)

Edmund Blunden writes more extensively on the mourning of the natural world in his poignant description of the river Ancre, which runs through Picardy in Northern France, eventually joining the river Somme. Written in 1925, the poem looks back on the war from some years' distance, yet the sound of the river still lingers in Edmund's memories, as a 'crying, … sad rill' with a 'troubling tone'. The final stanza evokes the pain of war, both for the human author trying to come to terms with the past and for the natural landscape, which still bears the scars of the conflict:

> The struggling Ancre had no part
> In these new hours of mine,
> And yet its stream ran through my heart;
> I heard it grieve and pine,
> As if its rainy tortured blood
> Had swirled into my own,
> When by its battered bank I stood
> And shared its wounded moan.
>
> Blunden, 'The Ancre at Hamel: Afterwards' (Silkin 1979: 107–08)

Here, as in the extract from Hasler above, the experiences of human beings and the landscape merge in an outpouring of pain and anguish. Likewise, in 'Report on Experience' Blunden interweaves his account of the destruction of human society with the loss of wildlife, just as the prophets Hosea and Jeremiah have done before him, noting the absence of birds and their prey as well as of people:

> I have seen a green country, useful to the race,
> knocked silly with guns and mines, its villages vanished,
> Even the last rat and last kestrel banished –
> God bless us all, this was peculiar grace.
>
> Blunden, 'Report on Experience' (Silkin 1979: 109–10)

Cosmic distress

The language used to depict the physical realities of warfare at times takes on an apocalyptic tone, such as Albert E. Tomlinson's reference, in 'Ghost of the Somme', to 'The sulphur stink, the black fumes, the blotted stars' (Copp 2001: 117). A number of poems go further than this and portray the conflict in terms of cosmic disorder, reminiscent of the biblical material. For A.E. Housman, newly recruited soldiers encounter a world that is shaking and crumbling with apocalyptic terror:

> These, in the day when heaven was falling,
> The hour when earth's foundations fled,
> Followed their mercenary calling
> And took their wages and are dead.
>
> Their shoulders held the sky suspended;
> They stood and earth's foundations stay;
> What God abandoned, these defended,
> And saved the sum of things for pay.
>
> Housman, 'Epitaph on an Army of Mercenaries' (Housman 1988: 138)

Both the use of the merism 'heaven and earth' and the notion that the earth has foundations like a building are typical of biblical descriptions of the world (e.g. Ps. 82.5, Isa. 24.18).

In what is probably one of the most evocative and poignant poems from the battle zone, Mary Borden depicts the anarchy and suffering of war in terms that she herself called 'fragments of a great confusion'. Yet lest we think that she has exaggerated her impressions, she adds 'to those who find them unbearably plain, I would say that I have blurred the bare horror of facts and softened the reality … because I was incapable of a nearer approach to truth' (2008: Preface). She writes:

> The sky long since has fallen from its dome.
> Terror let loose like a gigantic wind has torn it from the ceiling of the world,
> And it is flapping down in frantic shreds.
> The earth ages ago leaped screaming up out of the fastness of its ancient laws.
> There is no centre now to hold it down. It rolls and writhes, a shifting tortured thing, a floating mass of matter set adrift.
> And in between the fluttering tatters of the ruined sky,
> And the convulsions of the maddened earth,
> The man stands solid.
> Something holds him there.
>
> Borden, 'Unidentified' (Borden 2015: 35–6)

Borden's stark and striking choice of words conjures up for us a visual image of great force. As we have already seen (e.g. Jeremiah 4), the use of sensory language in the poem serves to heighten the emotional tenor of the description. But the act of seeing functions as more than a rhetorical device. Recent developments in trauma theory have highlighted the importance (and difficulty) of witnessing as part of making sense of an event (e.g. Laub 1995; Smith-Christopher 2011). Instinctively, both the Hebrew Bible and First World War poets act as witnesses to the events and so confront the trauma that they and their contemporaries face, leaving behind a lasting reminder of the devastation and futility of war.

Mother Earth

In her essay on British war poets, Trudi Tate notes that First World War authors frequently represent the landscape as 'mother', with war zones being remembered as the maternal body (Tate 2009: 171). This specific form of personification is largely absent from the Hebrew Bible war poems, where even the imagery of labour pains is applied to human suffering, as we have seen. However, Mother Earth is a frequent and important concept in Greek and Roman texts, suggesting that the war poets drew on classical sources as well as biblical ones in their work.[17] In Richard Aldington's graphic poem, the earth, personified as Demeter, the Greek goddess of agriculture, suffers devastating wounds with each shot fired. Here, implicitly at least, the soldiers' wounds are a result of the damage to Mother Earth:

> Each cruel bitter shriek of bullet
> That tears the wind like a blade,
> Each wound on the breast of earth,
> Of Demeter, our Mother,
> Wound us also,
>
> Aldington, 'In the Trenches' (Silkin 1979: 137)

But as well as protecting the troops, the earth is itself sometimes hostile, threatening to engulf or drown them. Isaac Rosenberg paints a disturbing picture of waggons rolling over dead bodies that are now in earth's grasp:

> Earth has waited for them,
> All the time of their growth
> Fretting for their decay:
> Now she has them at last! …
>
> Earth! have they gone into you!
> Somewhere they must have gone,
> And flung on your hard back
> Is their soul's sack
> Emptied of God-ancestralled essences.
>
> Rosenberg, 'Dead Man's Dump' (Bottomley and Harding 1977: 81)

Later in the same poem, Rosenberg represents the earth as both wounded victim and receiver of the dead, addressing 'Maniac Earth' as if she were the aggressor, yet depicting her as wounded with a 'bowel seared by the jagged fire'. He continues in this ambivalent vein:

> Dark Earth! dark Heavens! Swinging in chemic smoke,
> What dead are born when you kiss each soundless soul
> With thunder and lightening from your mined heart,
> Which man's self dug, and his blind fingers loosed?
>
> Rosenberg, 'Dead Man's Dump' (Bottomley and Harding 1977: 81)

Not all is doom and gloom, however. As we saw at the start of the essay, some poems hearken back to a golden pastoral era; others look forward to a future without war, where nature flourishes again. For example, the optimistic tone set by Charlotte Mew where trees are portrayed as old, wise women, patiently waiting for the spring to come and heal their wounds:

> Let us remember Spring will come again
> To the scorched, blackened woods, where all the wounded trees
> Wait, with their old wise patience for the heavenly rain,
> Sure of the sky: sure of the sea to send its healing breeze,
> Sure of the sun.
>
> Mew, 'May, 1915' (Kendall 2014: 46)

Comparisons and concluding remarks

A number of points of comparison between the two bodies of material have already been noted. It remains now to highlight some differences and explore the reasons behind them. Firstly, despite the examples cited above, explicit and detailed references to the destruction of nature in the First World War poems are fragmentary and rare. The majority of poems are concerned with human life and suffering, and allude to nature as a means of better articulating the horror experienced by the troops. By contrast the prophetic texts of the Hebrew Bible contain more frequent references and extended depictions of the effects of war (or other disasters) on the natural world. What is the reason for the apparent centrality of 'earth' language in biblical poetry when there are only snatches of it in the First World War poems? Throughout the Hebrew Bible, land or the earth[18] functions as a core theological theme, and the interplay between people and land under the watchful eye of YHWH is central. This is evidenced in a very particular understanding of cause and effect, as summed up in the blessings and curses of Deuteronomy 28. If the Israelites obey God, the land flourishes and prospers; if they disobey, it suffers. Unsurprisingly then, when the punishment for Israel's sinful behaviour results in enemy invasion as in Jeremiah 4, the land is also caught up in the conflict and suffers not just collateral damage but is part of the overarching theological narrative that governs the text.

A second, related issue concerns the subtle but significant differences between the way that Hebrew Bible prophets such as Jeremiah speak of the earth, and the First World War poets' use of language of the physical landscape. In many instances, the latter's personification of earth and the graphic description of its distress form merely a backdrop to the human story with all its horror. The environmental effects of war are depicted in passing in the First World War texts, and more often than not the fate of the earth seems to serve as a representation of human suffering and terror, rather than a primary concern of the poets. By contrast, the biblical authors genuinely attribute to earth a status in its own right as a character in the narrative of YHWH's dealings with the world. Key to this may be the biblical understanding that the whole world is created and owned by YHWH, and thus the relationship between the deity and his creation (human and non-human) is of paramount importance (Fretheim 2005). Although the First World War poets will also have been aware of this as part of their cultural and religious heritage, it does not take centre stage in that same way as it does in the biblical texts. In addition, the Hebrew Bible authors are motivated by their need to explain the catastrophe in theological terms as well as to describe it. As noted above, disasters, whether military or natural, are often attributed to YHWH's judgement on the Israelites for their disobedience. In this attempt to exonerate YHWH, the blame for both human suffering and ecological damage is laid at the feet of Israel's leaders (e.g. the 'shepherds' of Jer. 12.10–11).

This leads to an important question on the relationship between the two sets of texts: to what extent are the First World War poets drawing on biblical imagery and language, consciously or unconsciously? As previously noted there are some explicit references to biblical texts, for example the opening lines of Blunden's poem 'Report on Experience', in which he subverts Ps. 37.5 thus: 'I have been young, and now am not too old; And I have seen the righteous forsaken'.[19] There are also a number of 'theological' poems, mostly equating the noble sacrifice of the soldiers to Christ's death, and proclaiming the triumphant victory over evil that is to be their reward. Beyond these explicit references, how far are allusions to biblical themes and motifs, especially to the destruction of the earth, intentional? The social context of late Victorian and early Edwardian Britain was still highly religious and biblically literate, so biblical apocalyptic language, for example, would have been familiar. Moreover, these poets stand in a long tradition of literary works that retell biblical stories in graphic and passionate terms, from Bunyan to Milton and beyond. But this is also a period in which interest in the Classics was at its height and study of Graeco-Roman literature was considered essential to a good education. It is unsurprising then that Greek and Roman mythology also informs the language and ideas of many First World War poems, such that Elizabeth Vandiver can write 'the same classical tropes and images are cited, referenced and developed, to argue against the war effort but also for it' (Vandiver 2010: 5). For example, in her analysis of Rosenberg's Dead Man's Dump', Vandiver points out that the author movers seamlessly from biblical images to Homeric ones (Vandiver 2010: 299). While some authors undoubtedly consciously wove carefully crafted religious and classical themes into their poems, it is likely that many simply drew on the storehouse of myths, texts, phrases and themes that formed part of their cultural world.

It is perhaps only with hindsight, and with current scientific understanding of the on-going effects of ecological damage in the field of Northern France that we might read the poems that emerged from the First World War as a response to the effect of war on the land. For the Hebrew Bible poets, although they do not think in the same ecological categories as us, the damage inflicted on the physical landscape is much more in the foreground, tied up as it is with their theological understanding of the land as YHWH's gift and inheritance to them. Nevertheless, in both sets of texts, we find a merging of human and landscape in graphic accounts of the experience of war and its impact. In that respect then, both the ancient and the twentieth-century texts are ecological, representing attempts to remember, interpret and make sense of events that had a catastrophic effect, not just on human communities, but on the land that nurtured them, and that we also still depend on.

CHAPTER 10
PAGAN ANIMISM: A MODERN MYTH FOR A GREEN AGE
Ailsa Hunt

In a confrontational article that is now over fifty years old, Lynn White Jr went looking for 'the historical roots of our ecologic crisis' and found them, confidently pinning 'a huge burden of guilt' onto Christianity and its anthropocentric worldview (White 1967: 1206). White's five-page article – published in *Science* in 1967, and originally a lecture to the American Association for the Advancement of Science – is in many ways a maverick piece of scholarship, but it went on to have a seismic cross-disciplinary impact: White's influence can be strongly felt in environmental ethics, environmental philosophy, ecotheology, environmental history and ecocriticism.[1]

With medieval history his own disciplinary background, it was White's major interests in both the medieval Church and farming technology that led him to reflect on the relationship between Christianity, technology and nature, and the conclusions he came to were bold. As 'the most anthropocentric religion the world has seen', Christianity, when married with the emerging medieval technological capability, encouraged destructive treatment of the natural environment (White 1967: 1205). White's argument contains obvious flaws – such as sweeping generalizations and over-reliance on the idea of 'dominion' in Gen. 1.26–28, without once engaging with the text (even in translation) – as well as striking claims, and critics have been unsurprisingly swift to respond.[2] In particular, it has become commonplace for any scholars interested in the relationship between Christianity and our global environmental crisis to frame their thinking as a response to White.[3] Christian theologians led the way here – and in so doing established the field of Christian ecotheology – but by the beginning of the twenty-first century responses from within biblical studies had also picked up momentum.[4] Indeed despite often intense critique of White, these scholars frequently emphasize – and thus perpetuate – his impact: this 'generative thesis' (Jenkins 2009: 283) has 'created a virtual cottage industry among biblical scholars and theologians' (Tucker 1997: 3–4), earning its reputation as 'the most cited piece in the whole ecotheological debate' (Horrell 2010: 6.) For all its flaws, White's article raised questions to which many felt an urgent need to respond.[5]

White's claims about Christianity have thus been subjected to continuous criticism, but my chapter engages with an aspect of White's provocative article to which nobody (to my knowledge) has yet felt a need to respond. For White's claims about Christianity are not the only bold claims he makes about religious worldviews and attitudes to the environment. In fact, White holds up as an 'absolute contrast' to his portrait of Christian environmental thinking another image, this time of ancient pagan (by which he appears to mean Greek and Roman) thinking about the environment (1967: 1205).[6]

> In Antiquity every tree, every spring, every stream, every hill had its own *genius loci*, its guardian spirit. These spirits were accessible to men, but were very unlike men; centaurs, fauns, and mermaids show their ambivalence. Before one cut a tree, mined a mountain, or damned a brook, it was important to placate the spirit in charge of that particular situation, and to keep it placated. By destroying pagan animism, Christianity made it possible to exploit nature in a mood of indifference to the feelings of natural objects. (White 1967: 1205)

This characterization of Greek and Roman thinking about the natural world does not begin with White – as I will go on to show – but he does provide a particularly succinct and confident crystallization of the idea of 'pagan animism' and gives it an ecological agenda. This has also had its impact on academic discourse about our environmental crisis, but unlike White's claims about Christianity, an impact that has to date gone unrecognized and unchallenged.

My chapter aims to address this. I begin by exploring the origin and history of the idea that White terms 'pagan animism'. The idea that the Greeks and Romans had an animistic understanding of their environment has its roots in nineteenth-century comparativist scholarship and has been present in classical scholarship ever since, sometimes overtly and sometimes playing a much more muted role. I posit that White's depiction of the idea – and the official-sounding label 'pagan animism' – gave the idea new impetus and the ability to spread into other disciplines. I then argue why this is intellectually troubling: for pagan animism, as a way of characterizing ancient Greek and Roman theological responses to their environment, is flawed and unproductive. Yet the major focus of this chapter is not the ancient world itself but the history and power of an idea. This is a reflective piece about how the flawed idea of pagan animism has spilled out from classical scholarship into other disciplines, and there gets to work in shaping contemporary debate about fruitful responses to our environmental crisis.

I focus in this chapter on pagan animism's influence within Christian ecotheology and biblical studies, first because White has had a particularly visible impact within these fields, and second because narrowing the focus in this way allows us to engage in-depth with the material.[7] I put pressure on the intellectual impact of allowing this idea to travel across these disciplines without challenge. How has this flawed idea about the ancient world shaped the way Christian ecotheologians and biblical scholars construct theories of what a positive Christian environmental ethic might look like today? Thus this chapter asks – within a chosen set of boundaries – a much bigger question: how would thinking harder about the ancient world change contemporary environmental debate? As a warning tale about the power of an unchallenged idea, this chapter also functions as a microstudy of the book's wider agenda in exemplifying both the intellectual necessity and the productivity of cross-disciplinary dialogue in responses to our environmental crisis.

∞

White's claims about pagan animism are misguided. But it would be unfair, I will argue, to hold him solely to account for this. It is impossible to say who informed White's understanding of what Greek and Roman pagans thought about their environment, as his article retains the feel of a lecture in its lack of footnotes or academic references. *Quellenforschung* is therefore not advisable, especially given that White's reference to mermaids in his image of pagan animism hints strongly that he was not well acquainted with classical sources. However, my aim here is not to criticize White's research, as to show that it is understandable how he came to 'believe in' pagan animism. For scholars versed in Greek and Latin texts say similar – if rather more nuanced – things, and indeed had been doing so for a long time before White. Within classical scholarship there is a long history of linking animism, the environment and Greek and Roman religion, which I will now summarize.[8]

The study of both Greek and Roman religion as disciplines in their own right originated in the nineteenth century, when these defined areas of study emerged from an academic context in which research into religion typically adopted a comparativist and broad brush-stroke approach. The comparativists – as these nineteenth-century scholars are often called – were fascinated by origins. They expended much energy in seeking out the earliest seeds of religious thought, in order to produce theories, of worldwide application, which explained the emergence of religion. Among these, that of animism found particular favour.[9] Primitive thinkers – or so the animist argument went – interpreted the agency of the natural world around them as the presence of divine powers or spirits, which they then began to worship. Tylor's weighty *Primitive Culture* (1871) gave this theory the authority of established fact. It was then within this intellectual context that an increasing number of monographs devoted purely to Greek religion, and in their footsteps Roman religion, began to appear, with momentum picking up in the early twentieth century.[10] Scholars who chose to focus their attention on either Greek or Roman religion inherited from the nineteenth-century comparativists a lively interest in the origins of religion. Consequently, a driving aim of their research was to unveil primitive layers of Greek and Roman religion that they were convinced were waiting to be discovered. Moreover, because animism was viewed as the primitive form of religion par excellence, they turned to Greek, and in particular Latin literature, *expecting* to find views of the natural world which they could interpret in an animistic way. Unsurprisingly they found them. By far the most significant player here was the Latin word *numen*, understood to be a direct equivalent for *mana* (a Melanesian concept which the comparativists had held up as the quintessential example of an animistic worldview);[11] this strongly reinforced a tendency to present Roman religion as particularly primitive, vis-à-vis Greek religion.[12]

Today the word 'animism' is rarely heard on the lips of a scholar of Greek or Roman religion. Its reputation has been tainted by concerns about the imperialist and Christianocentric bias of the comparativists, as well as critique of their methodology, and focus has turned to other theoretical models, with that of *polis* religion enjoying particular influence.[13] And with the scholarly demise of animism, thinking about the environment by scholars of Greek and Roman religion has largely stagnated (something which the city-facing focus of *polis* religion has no doubt reinforced). Yet ironically, when the environment does receive any attention, we find the old animist orthodoxy reinforced, albeit in updated language.[14] Scholars of Greek and Roman religion today do not overtly endorse an animist position, but rather their discussions of Greek and Roman responses to the environment are replete with what I would call 'animist language': you can expect vague references to a 'divine aura' (Rives 2007: 91), the '*genius loci*' (Turcan 2000: 38), the 'sanctification of the landscape' (Turcan 2000: 38), 'naturally numinous places' (Mikalson 2010: 4), 'the presence of the gods in the natural world' (Rives 2007: 89), 'gods … present in the landscape' (Larson 2010: 57), 'awareness of the sacred in the landscape' (Larson 2010: 58), places 'charged with divine power' (Cole 2004: 64) or 'the sense of the divine (*numen*) inspired by nature' (Warrior 2006: 5). Rives sums up the thinking as follows:

> The natural world, then, would for many people have been shot through with the presence of the divine. By acknowledging this presence through offerings and other demonstrations of piety, they were able to bring the natural world and the powers that animated it into a defined relationship with the human sphere, while at the same time demonstrating their respect for those powers. (Rives 2007: 92)

Thus despite animism's general demise within scholarship on Greek and Roman religion, it persists when any attention is paid to the environment. It is important to note, however, just how rarely this happens: for many scholars of Greek and Roman religion the environment sits in a blindspot.[15]

Nor does the relationship of Greeks and Romans with their natural environment receive much attention outside of the fields of Greek and Roman religion. A handful of books explicitly tackle the relationship, questioning to what degree, if any, Greeks and Romans were aware of human-caused environmental problems, and investigating what kinds of attitude, both theoretical and practical, they might take to the natural world.[16] There is also a distinct lack of in-house expertise devoted to the study of Greek and Roman environmental thinking: Hughes, arguably the most prominent writer on this topic, is an environmental historian, not a specialist in ancient history. Yet within this niche area of classical scholarship, animistic ways of understanding Greek and Roman relationships with the environment are again easy to find. These scholars do not present the Greeks and Romans as 'environmental purists' in the way that White idolizes them: both Hughes and Thommen, for example, accept that the Greeks and Romans caused plenty of environmental damage.[17] But this is tempered by acknowledging that an animistic understanding of their natural environment meant that they would treat it with caution and care, with environmentally sound steps attributed to this animistic respect for nature. Hughes waxes lyrical on this theme.

> Animism in general provided an enchantment of nature that made people think twice before harming it. (2014: 65)

> The Greeks and Romans traditionally regarded the world as a sacred place where the gods of nature, who shared some human qualities, were present. Thus they were expected to treat the environment with awe and care, and this was practised to a considerable extent. (2014: 43)[18]

Particular interest attaches to sacred groves or sacred enclosures, where both Hughes and Thommen argue that environmentally friendly non-interventionist practices took place, thanks to a fear of damaging a place which had been specifically set apart for the divine to inhabit.[19]

Indeed the 'untouchability' of the natural world – with sacred groves being *particularly* untouchable – is also a common motif in scholarship on Greek and Roman religion: physical intervention with the land is said to require negotiation with the gods, although what this means in practice is often sidestepped. Turcan, for example, observes that 'all working of the land required an appeal to the gods' (2000: 40), whilst Cole writes that:

> In the countryside, any grove, spring, modest rock ledge, or cave could belong to a deity. Visitors were expected to understand that such places might be sacred and to realize that they should modify their behavior in a way appropriate to divine requirement. (2004: 40)[20]

With their priorities elsewhere, these scholars do not draw out the ecological impact of such thinking about the natural world, but their vague claims about a respectful, non-interventionist approach to the environment help to reinforce the conclusions of scholars like Hughes and Thommen who *are* preoccupied with environmental questions. And despite the vagueness of these claims, scholars following this line of thinking often write as though there must have been sets of rules structuring Greco-Roman interaction with their environment. Thus Hughes will observe that 'the rules protecting land reserved for the gods were strict, numerous, and followed a consistent pattern' (2014: 189) or gesture to 'a basic law found everywhere [which] forbade felling trees or cutting branches' (2014: 189), without specifying what those rules were, or where they are actually to be found.[21]

In short, the idea that Greeks and Romans engaged with their environment in an animistic way has a long history. White by no means invented the idea that he terms pagan animism and is not 'responsible' for it. Whilst he himself provides no evidence for his claims about pagan animism, he presumably picked up the belief that Greeks and Romans were animist in their approach to nature from earlier scholarship devoted to Greek and Roman religion or simply as part of 'received wisdom'. To repeat, White did not invent pagan animism. What he did do is as follows. One, he gave the idea an explicit ecological thrust: in this he was very much ahead of the curve. Two, he crystallized the idea in a bold and evocative image, giving it an official-sounding label and therefore the aura of authority.[22] Three, he gave the idea life outside of the world of classical scholarship. In 1967 White was a lone voice in claiming that Greeks and Romans took an environmentally friendly approach to the natural world thanks to an animistic understanding of it. Today, whilst the thinking of Greeks and Romans about the environment is still hugely under-studied, anyone turning to more recent classical scholarship to check the reliability of White's claims could take away the impression that Greeks and Romans engaged in environmentally sound practices because of animistic ways of thinking; anyone consulting works on environmental history, chiefly Hughes and Thommen, would take away the same message.[23] There is nothing here to put a hold on the new lease of life that White gave to pagan animism. And this, I argue, should worry us.

∞

The reason for worry is straightforward enough: for the idea of pagan animism is of doubtful value and insight when it comes to understanding Greek and Roman thinking about the religious dimension of their relationship with the environment, or in other words, about how theology and ecology intersect. This area of Greek and Roman thought represents a huge topic, and would be more than worthy of several book-length treatments. Consequently, I here take an illustrative approach, focusing on Greek and Roman thinking about trees, as a way of indicating the inadequacies of an animistic approach to the topic. I will highlight how an emphasis on animism has urged scholars from the nineteenth century onwards to agenda-led readings of ancient evidence for Greek and Roman relationships with trees, rather than dealing with the evidence on its own terms. I also reveal what we lose if we look at Greek and Roman religious engagement with their environment only through an animistic lens. Necessarily impressionistic, this section will point to a far wider picture of the richness of Greek and Roman theological thinking about their environment.[24]

There are several reasons for this arboreal choice (and not only that Roman thinking about trees has long been an interest of mine). First, attitudes to trees and deforestation feature prominently in modern environmental discourse, and may be treated as a kind of litmus test for a given community's broader environmental ethics.[25] More important, however, is the fact that trees are the 'go-to' example for anyone wanting to illustrate Greek and Roman animistic responses to the natural world. This was the case when animistic interpretations of religion had their heyday in the nineteenth century; it is the case in scholarship on Greek and Roman religion today; and it was the case for White as well.[26] The tree, and tree spirit which needs placating before it can be cut, enjoys first place in White's illustration of pagan animism (1967: 1205). Later he also enigmatically observes that: 'To a Christian a tree can be no more than a physical fact. The whole concept of the sacred grove is alien to Christianity' (1967: 1206). When characterizing pagan animism and its polar opposite Christianity, White believes that an attitude to trees and groves makes this opposition particularly clear; nor is he alone in seeing trees as a focal point of tension at which pagan and Christian attitudes to the environment clash.[27] In this section I focus on trees, as the favoured example of proponents of pagan animism, in order to debunk the idea.

Since the nineteenth century to the present, scholars exploring attitudes towards trees among the Greeks and Romans have cherry-picked the same handful of passages.

Whilst the earliest scholars discussed these boldly in terms of animism, tree spirits and tree worship, more recent scholars have leaned on less overt animistic phrases, such as 'frisson of the supernatural' (Turcan 2000: 39) or 'divine aura' (Rives 2007: 91). Here I consider three concepts that have been of particular influence: first, hamadryads (tree nymphs); second, the connection of groves with the word *numen*; third, prohibitions against tree felling. Whilst it is possible to see how the ancient evidence for these concepts *could* be used to bolster an animistic agenda, I will show that an attentive and contextualized reading of the evidence destabilizes this. I begin with hamadryads, who have been framed as proof of animistic attitudes in the Greek and Roman world, and of belief in a kind of 'tree soul'.[28] Of significance has been a depiction of nymphs in the *Homeric Hymn to Aphrodite* (in fact not specifically called hamadryads, but rather 'mountain nymphs', 257). The hymn depicts these nymphs as coeval with certain oaks and pines: at the nymphs' birth, the trees spring up, and when the trees die, the souls of both nymph and tree leave the light of the sun together (264–72). What goes unnoticed here, however, is that both have *separate* souls, undermining readings of these nymphs as the 'animating soul' of the tree. Another favoured passage is Ovid, *Metamorphoses* 8.725–76, where the impious Erysichthon fells an oak, and in so doing kills a nymph who cries out in agony from 'under this wood' (771). This passage has had an inordinate amount of influence, yet nobody stops to consider whether Ovid's presentation of Erysichthon as 'pantomime villain' should make us cautious of assuming widespread Roman objections to tree felling, due to an animistic fear of potentially harming tree spirits.[29] Attention has also turned to the number of Greek and Roman commentators (or scholiasts) who, leaning etymologically on *hama* meaning 'with' in Greek, define hamadryads as nymphs who are born and die with their trees (e.g. Servius, *Commentary on the Aeneid* 3.34).[30] Rather than providing proof of animistic arboreal attitudes, these passages, I argue, speak of the commentator's urge to define, clarify and clean up muddy waters. Unsurprisingly, full consideration of the literary depictions of hamadryads reveals a world of muddy and nuanced ways of imagining the relationships between hamadryads and trees.[31]

I turn now to two famous passages connecting the word *numen* with groves (Pliny, *Natural History* 12.3 and Ovid, *Fasti* 3.295–96). *Numen* is very challenging to translate. Much weight was attached to the word in early scholarship on Roman religion, when it was taken as a synonym for *mana*, and as proof of Roman belief in vaguely defined impersonal spirits (*numen* is a neuter noun) animating their natural surroundings. From the mid-twentieth century these views began to change, with some scholars choosing to understand *numen* as signifying the will or power of a deity.[32] However, animistic understandings of *numen* persist when it comes to features of the natural world. Thus when Ovid describes a shady grove under the Aventine (*Fast*. 3.295–6), telling the reader that they might think there was a *numen* within, this has frequently been interpreted as proof of Roman belief in impersonal spirits animating groves.[33] There are many problems with this, not least the extrapolation from a single example to a universal rule, and lack of consideration for context. The depiction of the grove sits within a mythological narrative of the exploits of Numa, Picus and Faunus (*Fast*. 3.289–322), in which the word *numina* (plural of *numen*) is three times used of Picus and Faunus, figures who are also described as gods and clearly act as anthropomorphic divine agents. In this context, it is hard to read Ovid's use of the word *numen* as unambiguously referring to an impersonal spirit. There is also nothing in this passage to suggest that Ovid uses *numen* in response to a perceived agency in the grove; rather it seems to be closely linked to the grove's shady and gloomy qualities. However *numen* is to be translated here, there is no prompt from the Ovidian context to understand it as an animating spirit of the grove. Meanwhile, at *Natural History* 12.3 Pliny writes that in ancient times trees were considered to be temples of *numina*, noting that in his day rural people still dedicated exceptionally tall trees to gods. He goes on to talk of tree-types being considered dedicated to their own *numina*, such as the oak to Jupiter, and the olive to Minerva. The word *numen* is clearly used here to refer to anthropomorphic deities, again destabilizing the possibility of reading the word, a few lines earlier, as referring to shadowy tree spirits.[34] In short, these passages have to be read with an animistic agenda in mind in order to be interpreted as proof of animistic thinking.

Scholars have also argued that Greek and Roman prohibitions against tree felling stem from animistic attitudes.[35] Much emphasis has been placed on an archaic inscription from Spoleto (*CIL*. 11.4766), which may forbid cutting of trees in a grove (depending on translation of the tricky Latin). Despite the fact that this is clearly a law for a specific grove, scholars have extrapolated universal rules about groves being inviolable, which they use to bolster arguments about the natural world's perceived animation.[36] But there is also ample evidence from the Greco-Roman world that felling trees was by no means assumed to be sacrilegious behaviour disrespecting the possible presence of a tree spirit, which I bring to attention in my monograph.[37] For example, Greek inscriptions from Lydia show us locals cutting down trees associated with a sanctuary of Zeus *until* an unexpected punishment warns them off; in the Arval grove a few miles outside of Rome, Latin inscriptions reveal a ritual pruning of the trees each May which was a mode of articulating their religious significance.[38]

In short, narrow focus on a handful of arboreal passages has blown them out of all proportion, with individual instances turned into universal rules. This focus has also blinded scholars to the colourful variety of ways in which trees had religious significance for Greek and Roman thinkers (and thus consolidates an image of pagan thinkers engaging with the arboreal world purely as animists). One famous Roman tree and one Greek can serve as examples: the Ruminal fig was considered sacred because of the memories it embodied as the tree under which Romulus and Remus were suckled by a she-wolf (Pliny, *HN*. 15.77), whilst the oak of Dodona was renowned for its ability to communicate messages from the divine through its rustling leaves (e.g. Homer, *Odyssey* 14.327–8; Silius Italicus, *Punic Wars* 3.680).[39] Indeed, understanding how trees might communicate the divine to a human audience was significant within Greek and Roman attempts to understand their position in the world.[40] Whole books were written on arboreal portents (as we know from Pliny, *HN*. 17.243). Trees also communicated ideas about the nature and identity of the gods. At Sparta Dionysus was known as Dionysus of the Fig, because the fig was his discovery (Athenaeus, *Deipnosophists* 3.78c), whilst at Delphi a hybrid Athena-cum-palm statue helped to visualize the goddess' relationship with victory (with which palm trees were strongly associated; Pausanias, 10.15.4; Plutarch, *Nicias* 13.3 & *On the Pythian Oracles* 397F).⁴¹ This quick roll call of examples sketches out how focusing on a handful of favoured 'animistic' passages is to miss out on the richness of the theological thinking prompted by trees in the Greek and Roman worlds.

To sum up, scholarly attempts to present Greek and Roman attitudes to trees as straightforwardly animistic are based on an agenda-led reading of precious little evidence, which blows that evidence out of all proportion. And in so doing these scholars ignore the rich variety of ways in which trees were of religious significance to the Greeks and Romans. The trees we have met here should warn us that taking the idea of pagan animism as a guide to understanding Greek and Roman attitudes to the environment will lead to unsatisfactorily thin and academically suspect conclusions. In the following section I explore what happens when White gives this flawed idea a new lease of life outside of classical scholarship.

The cross-disciplinary influence of White's image of pagan animism is most palpably felt when scholars repeat it without critical comment on its validity and value. This tends to happen when a scholar sums up White's claims, but because their focus in so doing is never on the strength of what he says about pagans (and frequently on the strength of what he says about Christianity), the picture of pagan animism is both allowed to pass unchallenged and reinforced by sheer repetition. Within environmental philosophy, for example, Attfield assures us that 'pagan animism involved respect for the guardian spirits of trees, streams, and hills', whereas 'Christianity allowed its adherents to disregard the feelings of natural objects' (1994: 22).⁴² The biblical scholar Horrell refers to White's characterization of Christianity as a force 'sweeping aside other ancient mythologies, with their cyclical views of time and their animistic sacralization of nature' (2010: 7). Sideris, much of whose work focuses on the intersection between religion and environmental ethics, refers to White's conviction that 'the animistic paganism that once envisioned powerful spirits inhabiting every realm of nature has gone by the wayside' (2006: 449). Within ecocriticism, Borlik refers to White's article as 'a provocative exposé on the consequences of Christianity's assault on pagan animism' (2011: 9). Acknowledging that the focus of these scholars is on summing up White's views, all appear to swallow the idea of pagan animism without question. This passive acceptance of pagan animism can further be seen when Rudd, a literary scholar and ecocritic, uses the idea (explicitly taken from White) as a building block within an argument. In an ecocritical exploration of 'being green' in late Medieval English literature, Rudd reads the ending of *Sir Gawain and the Green Knight* as a kind of enactment of White's depiction of 'spirits *in* natural objects [which] … evaporated' on the advent of Christianity (Rudd 2014: 35, quoting White): of the Green Knight's sudden exit from the romance she writes 'there is no clearer example of the spirit within natural objects disappearing in the face of Christian rejection' (Rudd 2014: 36).

It is clear that White helps to transmit an intellectually damaging orthodoxy about Greek and Roman engagement with the environment into many different disciplines, where it may get repeated, or used to build up an argument, without critical engagement with its validity. There are also more subtle ways in which White's image of pagan animism has influenced academic debate about our environmental crisis which I now explore, with my focus, as promised, on responses to the crisis within biblical studies and ecotheology. White's distinction between environmentally friendly pagan animists and anthropocentric Christians easily made an impact within these disciplines because it chimes with pre-existing deep-seated ways of thinking about the history of Christianity. Significant here has been an emphasis on Christianity as a religion revealed in *history*, and the Bible as the story of God's saving history with humanity, which has undermined the relevance of nature to Christian belief and practice.[43] Such thinking gained momentum in the Reformation, when a new focus on the centrality of human salvation either sidelined thinking about creation, or emphasized its 'fallen' status.[44] This way of thinking then flourished in the biblical theology movement which developed in the mid-twentieth century, when an understanding that 'Christian theology has to do primarily with human history … not with nature' became commonplace (Santmire 1985: 4). Nature had a subsidiary role to history as the backdrop against which God acted, indeed a 'handmaiden' or 'servant of history' (Wright 1952: 43).

Scholars of this period were also influenced by Hegel and his idea that religious thinking progressed from a primitive stage until it reached the pinnacle that was Christianity; ancient Israelite religion – or Jewish religion, as Hegel called it – represented a step on from 'immediate religion' or 'nature religion' (which was the 'lowest level'), with Jewish thinking framed as an 'elevation of the spiritual above the natural'.[45] Drawing a contrast between Old Testament or Israelite views of nature and those of surrounding religions thus became a key feature of this scholarship. Wright, both a biblical scholar and an expert in archaeology of the Ancient Near East, sharply distinguished ancient pagan polytheists, focused on the 'rhythmic cycle of nature', from 'biblical man', focused on God's 'historical acts' and 'historical promises' (1952: 24–5).[46] On this distinction in outlook, the book of Genesis was felt to be particularly informative. It was argued that Genesis accounts of God creating the world deliberately separated God from nature, a defiant gesture of self-identification in opposition to surrounding polytheistic religious beliefs; these were not well-defined, but might loosely be referred to as 'nature-religions'.[47] In *The Secular City*, published in the same year as White's article, Cox presented Hebrew thought – which 'separates nature from God and distinguishes man from nature' (1967: 22) – in stark contrast to the thinking of those whose 'glens and groves swarm with spirits' (1967: 21). Later, Christian theologian Jürgen Moltmann also summed up this influential idea as follows:

> Because it grew up in an environment moulded by pantheistic, matriarchal, animist religions, the belief in Yahweh to which the Old Testament testifies, laboriously and perseveringly taught *the difference* between God and the world. (1985: 13)

The religions against which Hebrew thought self-identifies are either explicitly depicted as animist, or implied to be, swarming as they are with spirits. A stark distinction is drawn between the God of Genesis, who creates nature and is wholly separate from it, and 'nature religions' or 'animist religions', where the focus of worship and the natural world blur.

To sum up, White's contrast between pagan animists and Christians in their attitudes towards nature would not have shocked Christian theologians and biblical scholars used to thinking of the history of Christianity as a religion which deliberately turns away from nature: after all, White was speaking to the world of the biblical theology movement. White's distinction in fact reassuringly replicates the distinction drawn between the God of Genesis, who is separate from nature, and the worldviews of surrounding 'animist religions'. Fittingly enough in a religious tradition often thought of as 'Judaeo-Christian', the arrival of Christianity and its impact on pagan animism can be seen to mirror the arrival of Israelite religion and its impact on neighbouring religious cultures. What White did do, however, was to help turn this contrast from a positive to a negative: here the shock factor is undeniable. With the environmental awakening of the 1960s to 1970s, the commonplace that Judaeo-Christian thinking takes a non-reverential attitude to the environment could no longer be framed as a point for praise, and White makes this painfully clear.[48]

The challenge was now to respond. In what follows I single out three Christian responses to the environmental crisis in which unease about promoting an animist-sounding response is palpable, in order to illustrate how pagan animism has shaped them. I begin with biblical scholar Richard Bauckham and his book *Bible and Ecology*, which sets out, for a general Christian audience, a biblical vision of the human place within a 'community of creation'. Bauckham argues that in biblical thought 'nature is certainly de-divinised but it is not de-sacralised' (Bauckham 2010: 86): in other words, nature is not equated with God or the divine, but this does not deprive it of a sacred quality, or an ability to reflect the nature of God; and this in turn will – to put it simply – encourage a more respectful environmental ethic. Bauckham takes a careful and conciliatory approach, trying to bridge a perceived gulf between biblical teaching and a reverential approach to the environment. For example, Bauckham draws attention to Psalm 148, which he believes demonstrates that nature is not 'de-sacralised' in Christian thought. He presents this psalm as an excellent antidote to the anthropocentrism of which White accuses Christianity, encouraging instead a theocentric sense of the community of creation.[49] However, he acknowledges that many Christians may struggle with the psalm's picture of all creation worshipping God as 'they may take it to reflect some kind of pre-scientific animism or pan-psychism that attributes rational consciousness to all things' (2010: 79). Here Bauckham reveals a concern that if he pushes Christians too hard towards a reverential approach to nature, an instinctive fear of acting like an animist will kick in.

What Bauckham works to avoid can be illustrated by briefly turning to the work of Sheldrake, a biologist known for his controversial theories (especially that of 'morphic resonance'), who also identifies as Christian. In *The Rebirth of Nature* (1990), Sheldrake argues – contra Bauckham – that nature *has* been 'desacralised' in Christian thought (1990: 145). What Christian thinking now needs, he continues, is a 'rediscovery of the God of the living world' and a major way to achieve this is through a 'revival of the animistic traditions which prevailed until the Protestant reformation' (1990: 153–4).[50] Sheldrake does not argue that Christian history is completely lacking in animistic understanding, but that today it is barren of such thinking, sending many Westerners to explore religious traditions such as Hinduism or Buddhism, or 'to revive aspects of pre-Christian paganism' (1990: 154).[51] His take-home message is that only a move 'from humanism to animism' will make a 'resacralization of nature' possible (1990: 173). Sheldrake's views have been received with no little controversy. (Lovelock's review, in *Nature*, whilst warm overall, still felt a need to assure his audience that the book wasn't 'dangerous', although aspects of it may be 'nonsense' (1990: 685); and in a *Guardian* review Schwarz noted a likely disconnect between Sheldrake and Christian communities (1991: 33)).[52] Calls to 'resacralise' nature by 'becoming animists' are unlikely to cut the mustard in Christian communities on the ground, something of which Bauckham is acutely aware, as he treads a careful balance between a desire to restore reverence for nature and a concern about pushing his audience to readings of biblical texts which they may perceive as animistic.[53]

Unease about appearing animistic is also felt in the work of two influential theologians who have worked to reinvigorate the doctrine of creation in Christian thought: Jürgen Moltmann and Sallie McFague. Moltmann's arguments in *God in Creation* build on the doctrine of the trinity – the much-debated idea of God as three persons – to present an image of God who is both creator of the natural world, and thus transcendent over it, but is also very much present, indeed immanent, within it. This new thinking is about a 'recognition of the presence of God *in* the world and the presence of the world *in* God', which privileges vocabulary such as 'indwelling' 'sym-pathizing' and 'participating' (1985: 13–14). And it is, Moltmann argues, through the person of the Spirit that God indwells in the natural world.

> Creation exists in the Spirit, is moulded by the Son and is created by the Father. It is therefore from God, through God and in God. The trinitarian concept of creation binds together God's transcendence and his immanence. The one-sided stress on God's transcendence in relation to the world led to deism, as with Newton. The one-sided stress on God's immanence in the world led to pantheism, as with Spinoza. The trinitarian concept of creation integrates the elements of truth in monotheism and pantheism. In the panentheistic view, God, having created the world, also dwells in it, and conversely the world which he has created exists in him. This is a concept which can really only be thought and described in trinitarian terms. (1985: 98)

Moltmann argues that his trinitarian mode of thinking allows him to tread a delicate line between an over-emphasis on God's transcendence, on the one hand, and a pantheistic concept of God on the other; he also presents the latter as 'animist notions', with the two terms 'pantheism' and 'animism' treated fairly indistinguishably (1985: 99). The idea of perichoresis – a term used of the interpenetration of the trinity's three persons – enables Moltmann to construct a God who preserves his traditional role as creator of the universe, whilst being 'at one' with his creation. And this opens up the way for a Christian environmental ethic which must be highly attuned to the suffering of the earth, as well as the role it will eventually play in the new creation, which is a prominent theme in Moltmann's thinking. Moltmann insists that this conception of the person of the Spirit is neither animism nor pantheism, but rather pan*en*theism: 'all in God', rather than all being God, which allows his God and the universe to remain ontologically distinct. As he works to promote traditional theological grounds for a Christian 'reverence for nature', Moltmann relies on the concept of panentheism to insist that he has not inadvertently slipped into animist thinking.

McFague also works to give her 'ecological theology' a traditional point of departure, noting that Christianity is 'the religion of the incarnation *par excellence*' (1993: 14). As an eco-feminist, she starts from the understanding that forms of oppression – particularly that of women and nature – are interconnected, and sees questions about the importance and meaning of 'body' as a crucial link between the two.[54] Body is thus central to McFague's theology, in a way that it is not for Moltmann. Yet despite different driving agendas she takes a conceptual approach to the universe which is not unrelated to Moltmann's, if more radical, when she poses her opening and challenging question: 'what if we dared to think of our planet and indeed the entire universe as the body of God?' (1993: 19).[55] Like Moltmann, McFague also treads a delicate line between emphasis on God's transcendence and his immanence, arguing that God is 'the inspirited body of the universe, the animated, living spirit that produces, guides, and saves all that is' (1993: 20). Aware of a 'highly ambivalent' attitude towards pantheism within Christian thinking, she too leans on the notion of panentheism, which allows her to imagine God both in an agential mode (i.e. transcendent) and organic mode (i.e. immanent) (1993: 140).[56] Her model of God is a 'combined agential-organic one of the universe (world) as God's body, a body enlivened and empowered by the divine spirit' (1993: 142).

Both Moltmann and McFague rely on the term panentheism to ensure that they are not seen as pantheists or animists, whilst at the same time working to 'remerge' God and the world, in an ecologically motivated response to the long-reigning orthodoxy that Christian belief demands a non-reverential approach to nature. The language that both use to describe their models of God is animistic in tone ('indwelling'; 'sympathizing'; 'enlivened'; 'inspirited body'; 'animated, living spirit'; 'suffusing and energizing'); their models also foreground spirit – for Moltmann, the Holy Spirit and person of the trinity, but for McFague 'the divine spirit', or spirit of God who hovers over the waters in Gen. 1:2.[57] Perhaps these factors heightened the authors' unease about appearing overly animistic. Yet this unease is despite the traditional trinitarian element to Moltmann's ecotheology, and despite the fact that their arguments about God's immanence in the world (via the Spirit) have biblical roots in Pauline depictions of Christ being immanent in the world (Eph. 4.10, Col. 1.17), who can also be described as 'life-giving spirit' (1 Cor. 15:45). It is also despite the fact that their use of animistic language is situated in a monotheistic context, and the way such terms are understood is a far cry conceptually from the idea of a world populated by spirits of trees and springs. In short, whilst Bauckham, Moltmann and McFague construct varyingly radical responses to our environmental crisis (McFague arguably does divinize the world, something which Bauckham adamantly avoids), all are highly uneasy about running up against a deep-seated conviction in many Christians that they are fundamentally 'non-animists'.

This self-perception among Christians owes much to traditional thinking about Christianity having nothing to do with nature. White's stark dichotomy between pagan animists and Christians has no doubt also cemented the idea of animists as 'other', first in academic circles, and then affecting Christian communities who engage with academic calls to environmental action. I also suggest that popular perceptions of neopagans from outside the neopagan community – where a common stereotype may be that of a robed and bearded man communing with a tree-spirit while the bulldozers move in – reinforce a Christian sense of being *by definition* non-animist. Knee-jerk and insensitive as these perceptions might be, it is nevertheless true that many neopagans identify as animists or acknowledge animistic beliefs, something which they also persistently link to their environmental beliefs and practice.[58] As the contemporary animist Harris puts it, 'the organic environment is sacred to most Pagans and they are almost universally animist' (2013: 404).[59] Within neopaganism there are many kinds of animistic belief, yet a consistent link is the idea that animism can help to achieve ecological renewal.[60] The prominent Wiccan Lamond, for example, argues that our best chance of ecological survival is 'a return to the pantheist polytheist religious paradigm of our ancestors' (1997: 141).[61] Neopagans also often present their religious views, especially those concerning nature, as starkly at odds with Christian views. Christian environmental attitudes are seen as the central enemy in the pagan fight against ecological apathy and destructiveness: Lamond, again, blames 'the Protestant West's inventiveness' for 'spewing out pollutants' into our oceans and atmosphere (thus linking technological advances and Christian attitudes in a way strongly reminiscent of White) (1997: 29).[62] Moreover, it is neopaganism's *animism* which is said to set it so at odds with Christian thinking, which entirely separates God from nature. As York puts it:

> this comprehension of nature's inherent vitality or fundamental animism is what distinguishes paganism or paganisms from non-pagan transcendental or gnostic religions such as the God-is-wholly-other comprehension of Christianity and Judaism. (2003: 46)[63]

White's message that where Christianity is an ecological failure, pagan animism has the answers, is lived out in the thinking of today's neopagan animists. And from the Christian perspective, this only reinforces the perceived barrier to adopting an environmental ethic that in any way smacks of animism.

In short, there is a long-standing history of Christians thinking of themselves as diametrically opposed to nature-worshipping, animistic pagan polytheists. By the 1960s, however, such ways of thinking about Christianity's relationship with nature had become deeply problematic with respect to our ecological situation, as White made inescapably clear. Bauckham, Moltmann and McFague have provided examples of scholars who responded with creative and challenging ways of reimagining Christian relationships with nature, and of reinvigorating the doctrine of creation, whilst striving *not* to look like animists in the process.

∞

This chapter has worked to undermine the power of an idea. I have argued that the idea of pagan animism is born of a nineteenth-century assumption, is driven by an agenda-led reading of a handful of Greek and Latin texts, and has no solid evidential basis; uncritical thinking perpetuates its power. We have also seen this flawed idea spill out from classical scholarship into other disciplines, via White, who gives it an environmental agenda. The story of pagan animism and its influence on environmental debate teaches us that those driving that debate today ignore the ancient world at their peril. Ideas about the ancient world do shape contemporary theories of how best to respond to our overwhelming environmental challenges. Yet whilst the idea of pagan animism has had a huge impact, it has ironically received little attention. This chapter has rectified this by subjecting the idea to hard critical attention and has, I hope, seriously undermined its validity and worth. To challenge this idea would not be possible without critical introspection within a discipline, and to challenge its reach would not be possible without cross-disciplinary dialogue, the necessity of which this volume illustrates. But what difference does it make to have cut pagan animism down to size? Or in more general terms, how does thinking harder about the ancient world change contemporary environmental debate? By way of conclusion, I set out three ways in which undermining the idea of pagan animism should change modern environmental debate. My conclusions naturally focus on environmental debate within ecotheology and biblical studies, but further implications for other disciplines are as wide as White's cross-disciplinary reach.

The first change is straightforward. To dismiss the validity of pagan animism should halt the spread of this flawed idea and put an end to its being used in arguments, say, about Christian environmental ethics, or in ecocritical readings of literary texts. Cross-disciplinary reflection is here about ensuring accuracy, academic integrity and fairness to the ancient evidence. The second is as follows: if pagan animism is de-validated as a concept, then this pulls the rug from under White's dichotomy between Greek and Roman pagan animists and anthropocentric Christians. Ecotheologians, and biblical scholars engaged in environmental debate, could be far less uneasy about animistic elements in their environmental responses, if (at least some of) the pagan polytheists against whom they have traditionally self-defined are in fact not animists.[64] Animism need no longer be seen as something quintessentially pagan, other, non-Christian. Third, I wonder whether this disassociation of ancient paganism and animism might open up dialogue between Christians and neopagans in response to our shared environmental crisis.[65] If ancient pagan animism no longer stands as a polar opposite to Christian environmental behaviour, perhaps attitudes towards neopagan forms of animism might soften. At the academic level, there appears to be much on which neopagan ecoactivists and Christian ecotheologians might agree. York's criticism of a Christian transcendental view of God, which encourages ecologically destructive behaviour, is exactly the kind of accusation which theologians like Moltmann and McFague are working to escape, proposing more ecologically fruitful models of conceiving God's power and relationship to the earth. Reducing Christian opposition to the idea of animism might enable such Christian ecotheologians to work more closely with neopagan counterparts, without fear of backlash from the Christians in the pews whom they hope will live out their theories, as well as encouraging practical collaboration between Christian and neopagan communities.

Wondering where pagan animism's demise could take Christian ecotheology and ecoactivism is of course theoretical. But my main conclusion is not. The idea that Greeks and Romans took a non-interventionist approach to the environment because they were animists is a myth. It is a myth that White makes intensely appealing for a green age. In his hands, it gives us a glimpse of a world which is attractively eco-friendly and a dichotomy which explains why modern Western cultures, profoundly influenced by Christian thought, are so far from it: but it is nevertheless a myth, and one which should not be allowed to shape contemporary environmental thinking any longer.

NOTES

Introduction

1. The term 'Anthropocene' is not without controversy, with some geologists arguing that only clear evidence of changes in rock strata can support such a change in geological time frames. Others maintain that evidence of the domination of *homo sapiens* will eventually be detectable in the fossil record, as have other major changes that distinguished previous epochs (Kunkel 2017). For a fuller discussion see Williams et al. 2011.
2. In 2005 an estimated 84 per cent of the world's population declared themselves to be adherents of a religious tradition (http://www.adherents.com/Religions_By_Adherents.html, accessed 29 May 2018).
3. The term 'ecological hermeneutics' was coined by the Earth Bible Project, led by Professor Norman Habel, in 2000. The project adopts a very specific methodological approach to reading biblical texts ecologically, that differs considerably from the triangular model outlined here (see Habel 2000; Marlow 2009: 86–95).
4. While individual advances have recently been made in theorizing how ancient texts might inform modern responses to ecological issues (e.g. Horrell 2010; Lane 2011; Marlow 2009), this volume provides much needed cross-disciplinary discussion of such an approach.
5. This potential is nicely illustrated by the blend of religious and cultural influences informing the *Sibylline Oracles*, a text showcased in Van Noorden's chapter.
6. See Thommen 2012: 1–3 for an overview of the kinds of environmental damage inflicted by Greeks and Romans.
7. From a presentation Speth gave in 2006 at a retreat for US church leaders and scientists, hosted by the National Association of Evangelicals and Harvard University's Center for Health and the Global Environment.
8. Lane's approach is similar to that of ecocriticism in Literary Studies, which emphasizes that a reader approaches a text from their own political and moral orientation, and hence ecocritical readings often explicitly include a challenge to contemporary environmental ethics. Michael Branch suggests that ecocriticism is a call for cultural change and transformation, 'a move towards a more biocentric world-view, an extension of ethics, a broadening of humans' conception of global community to include nonhuman life forms and the physical environment' (Branch 1998: xiii).
9. See Hinds 1998 for one influential take on this huge subject area.
10. We are grateful to Nicholas Denyer for discussion and input on ancient philosophy.

Chapter 1

1. As Joseph D. Countryman, formerly the head of reservoir operations for the Army Corps of Engineers, said about the risk of flooding at the Oroville Dam in February 2017, 'When you build a dam, you are playing God … and it's tough to be God' (Nagourney and Fountain 2017).
2. Comparison between the Greeks and the Hebrews is a central part of discussions of the Axial Age, on which see the volumes edited by Eisenstadt 1986 and Árnason et al. 2005.
3. The lecture was given on 26 December 1966 at a meeting of the American Association for the Advancement of Science and first published in the 10 March 1967 issue of *Science* 155 (1967): 1203–07; it is cited here from the reprint in White 1974. I thank Rob Socolow for calling this article to my attention. For comparison, see the less pessimistic narrative account of Passmore 1974. White's legacy is the focus of Hunt's chapter in this volume.
4. For a comparison between the Hebrews and Greeks that touches on nature, see Boman 1960, although his section on 'History and Nature' (168–73) is more about cosmology than about nature in the environmental sense. For a further comparison centred on the two peoples' respective views of nature, see Manning 2010. 'Ecology' is not a transliteration of a Greek word; it was coined in German as *Ökologie* by the biologist Ernst Haeckel in 1866, from the Greek roots *oikos* (meaning household, with a sense of household economy) and *logos*. See Hughes 1994: 4 with n.8, and Tress 1997: 33, both citing Haeckel 1866; see also Thommen 2012: 6 with n. 4 for more discussion.
5. See Schofield 2006: 30–5, discussing this sense of *politeia* as a way of life in the surviving fragments of Critias' prose and verse writings titled *Politeia of the Spartans*; Xenophon's prose work of the same name; and Hippodamus of Miletus' account of the 'best *politeia*'.
6. Compare Thoreau's remark on huckleberrying as a form of relationship with the divine, which gifts us with a part of Eden that it is innocent for us to continue to enjoy: '[Berrying] is a sort of sacrament, a communion, – the *not* forbidden fruits which no serpent tempts us to eat.' See Thoreau 2000: 52, discussed in Lane 2009: 355–8.
7. There is however a 'personal is political' form of politics in the gendered relationship between Adam and Eve, which is beyond my scope here.
8. Hesiod, *Works & Days*, in West 1978, including the partially attested line 120; translations of this text are taken from Schlegel and Weinfeld 2006.
9. Baldry 1952: 91, argues that it was precisely 'to explain the present condition of man, particularly his need to work', whereas once he had 'lived on nature's bounty without toil', that Hesiod borrowed or invented the idea that the mythic age of Kronos was associated with gold (rather than, as was more traditionally believed by the Greeks, gold and other precious metals being a threat to its simplicity). Baldry 1952: 86 also notes the fragments of the Greek comic poets in whom we find a 'desire to caricature the traditional belief [in spontaneously available food] by carrying it to absurd lengths – in particular, the idea of

Nature producing her fruits of her own accord' exemplified in 'Telecleides' rivers of soup and self-frying fish, and Crates' self-moving gadgets which will end the need for slaves'.
10. For a rich discussion of the age of Kronos motif before Hesiod, in his poetry, in Old Comedy – focusing on the ideal of the *bios automatos* (the spontaneous production of life's necessities by nature) – and Plato's critical appropriations of these precursors, see El Murr 2009.
11. Translations of Plato are taken from those collected in Cooper 1997; the Greek of Platonic texts is cited from the most recent Oxford Classical Text.
12. Hillel 2007: 243; this quotation features in Appendix 2 (242–6), 'Perceptions of Humanity's Role on God's Earth', which seeks more broadly to refute Lynn White's assigning primary responsibility for ecological crisis to the Hebrew Bible and biblical religion (White 1974).
13. Contrast Hillel's assertion (2007: 245), that it is only with the expulsion from Eden that toil becomes necessary to produce food: 'The expulsion from the Garden of Eden is a folk memory of the beginning of agriculture. With that transition, humans no longer dwelled idyllically in parkland, feeding on wild fruits or animals, but had begun the toilsome cultivation of cereals.'
14. For example, Hughes 1994: 73, and Glacken 1967: 121, both cited in Goldin 1997.
15. Goldin (1997: 77–8) is more certain about this implication being present in the text than I am.
16. All translations from the Bible are from the Jewish Publication Society, as reprinted in The Rabbinical Assembly 2001.
17. Boman (1960: 173) argues that 'the belief in creation' arose out of the revelation of God's role in the deliverance from Egypt. Terence E. Fretheim treats God's role in the Noahide flood (Genesis 6–9) as exemplary of the subtle interplay between human and divine factors in natural outcomes that can be viewed as judgments more generally. He writes: 'In sum, people's sin generates snowballing effects. At the same time, God is active in the interplay of human sinful actions and their effects, and "third parties" are used by God as agents for that judgment (e.g. flood waters; Babylon). Both divine and creaturely factors are interwoven to produce the judgmental result.' He notes further that the Hebrew words for 'corruption' and 'destruction' in Gen. 6.11–13 'have the same root (*šḥt*)', showing that 'the judgment flows out of human wickedness, referencing the appropriate functioning of the created moral order', and can elsewhere (Hos. 4.1–3) have 'disastrous ecological effects' without further divine intervention. Indeed Fretheim observes that 'God is the author of a specific natural event only in Gen. 8:1, where God makes a salvific move, making a wind blow over the face of the waters' (Fretheim 2012: 700).
18. Ancient Atlantis too is ascribed autochthonous origins at *Criti*.113c.
19. For a useful discussion, see Kasimis 2013: 3–4, with the works she cites there by Vincent Rosivach and Nicole Loraux.
20. Compare the way that Montaigne in his 'Apology for Raymond Sebond' would deny that nature was an unjust stepmother, meaning that she provides for us adequately, while

Rousseau in his *Reveries of a Solitary Walker* affirms that she is a true mother indeed. For discussion and citation, see Lane 2009: 353 with n. 65.

Chapter 2

1. Lightfoot (2007: 203–53) explores this question, showing, for instance, how Sibylline versions of events 'track' particular Biblical prophetic expositions, while the use of Homer and Classical accounts appears more often to stay on the level of individual phrases or ideas.
2. Unless otherwise specified, translations of passages from the *Sibylline Oracles* are from Collins 1983, unless the passage is from books 1–2, for which Lightfoot 2007 is used.
3. All translation of the *Eclogues* are from Lee 1984.
4. The phrase *kata gaian* (*Sib. Or.* 3.491) might mean 'as regards the earth' (so Collins 1983 *ad loc.* – although rare, there is precedent in Demosthenes) or 'on the earth' (Buitenwerf 2003) but in context it seems more likely to mean 'throughout the earth' (by analogy with *kata polin*) or even the distributive sense 'land by land' (by analogy with *kat' ethne*), since it introduces a series of 'weal and woe' oracles directed at different cities and races.
5. The idea of paradise on earth as a *cultivated* space has been described as a particularly Jewish note; for a survey of Biblical passages to support the idea that wilderness is to be equated with God's curse, see e.g. Nash ([1967] 2001: 14–17) and certainly this space is not desirable in the *Sibylline Oracles*. However, the wilderness is in both the Hebrew and Christian scriptures also the site of encounter with the divine; see further Marlow (2009: 174–7).
6. Compare Hogeterp, A. on the *Dead Sea Scrolls*, in Marlow et al. (forthcoming).
7. I am grateful to Will Brockliss, Assistant Professor of Classics at the University of Wisconsin-Madison, for prompting thought on this point.
8. The destruction referred to might be Pompey's conquest of the temple in 63 BCE but the event is not mentioned anywhere else in the work; more likely it is envisaged as an eschatological event.
9. On the deliberate obscurity of apocalyptic images, see Sandy 2002: 34–96 and Manietti 1997: 246.
10. It should be noted, however, that Gauger's ([1998] 2011) edition of the text offers a slightly different reading of the Greek, such that there is still a personal agent of destruction.
11. See Lane in this volume, pp. 13–23 for discussion of the biblical plagues as 'signs'.

Chapter 3

1. Plutarch, *On the Self-Contradictions of the Stoics* 1052c, *Stoicorum Veterum Fragmenta* 2.604. All translations in this chapter are my own.
2. As Marcus himself reminds us at *Med.* 11.1.2.

3. See e.g. Gregory 2007.
4. See further, Sedley 2007: 31–61.
5. The fundamental idea had been anticipated by Herbert Spencer, who also coined the term 'super-organic' (Spencer 1876).
6. See Sedley 2016, on which this part of the chapter is based. The text numbers (B27, etc.) refer to fragment numbers in the standard collection of Presocratic philosophers, Diels-Kranz 1952.
7. See esp. *Med.* 10.7.2; cf. 5.13.
8. These calculations, variants of which are often quoted, are of course both approximate and subject to regular revision. Up-to-date data are available from the Ecological Footprint Network, http://data.footprintnetwork.org/#/. Cf. also http://www.bbc.co.uk/news/magazine-33133712.

Chapter 4

1. Translations of Seneca's *Letters* are my own.
2. 'Ecosophy T' is the name of the position advocated by Arne Naess, one of the founders of the Deep Ecology movement. In his own words (2003: 271–2), 'Ecosophy T has only one ultimate norm: "Self-realization!" I do not use this expression in any narrow, individualistic sense. I want to give it an expanded meaning based on the distinction between a large comprehensive Self and narrow egoistic self as conceived of in certain Eastern traditions of *atman*. This large comprehensive Self (with a capital "S") embraces all the life forms on the planet (and elsewhere?) together with their individual selves (*jivas*). If I were to express this ultimate norm in a few words, I would say: "Maximize (long-range, universal) Self-realization!"'
3. Much acclaimed, the so-called Lynn White thesis is still widely discussed (see e.g. LeVasseur and Petersen 2017; Jedan 2017).
4. Different authors have pointed out that the contrast between the Christian West and other non- or pre-Christian cultures is unconvincing, given the evidence in the latter for exploitative attitudes towards nature (e.g. Taylor 2010: 12). For a general evaluation of the Lynn White thesis, see Jedan 2017.
5. Individual Stoics: Posidonius (Protopapadakis 2012); Posidonius and Seneca (Dragona-Monachou 2015); Marcus Aurelius (Dragona-Monachou 2010).
6. On Stoicism and environmental ethics, see Wildberger's chapter in this volume, pp. 63–74.
7. See also *Nat. D.* 2.78 on human beings as junior partners in the cosmic city.
8. Contrary to what some authors claim (see e.g. Plumwood 1993), this need not be a bad thing. Granted, we are bound to be disappointed if we misidentify the adequate attitudes towards nature with a radical demotion of human reason. Demoting human reason, however, paradoxically undermines the very agent to whom ethics must appeal for any process of attitudinal change.

Notes to pp. 56–64

9. In so doing, I arrive at a far more positive interpretation than is found elsewhere (e.g. Sedley 2007: 205–38).
10. See Plutarch, *On Stoic Self-Contradictions* 1051F–1052B.
11. See also Seneca, *Natural Questions* 3.30.
12. See Diogenes Laertius 7.104–5 = LS 58B.
13. See LS 58C.
14. See Epictetus, *Discourses* 2.6.9 (= LS 58J). Cf. also Cicero, *On Duties* 3.13 and my interpretation in Jedan 2009: 121–42.
15. In recent discussions, Michael Slote has presented such a strong version under the label 'agent-based virtue ethics' (e.g. Slote 1995). The difference lies in Slote's developing his strong version as alternative to eudaemonist forms of virtue ethics, whereas eudemonism was the unquestioned point of departure for the Stoics.
16. Cf. e.g. Attfield 2014: 48–9.
17. For a Platonic voice on the threat of negligibility, see Lane 2011.
18. *Letters* 95.19 is by no means unique in Seneca's works, see e.g. *Natural Questions* 5.15 on mining: 'Those famous ancestors of ours, whom we are always heaping with praises, whom we complain that we do not resemble, cut down mountains, lured as they were by hope, and stood there over their profit – but under a mass of rubble.' (5.15.2, trans. Corcoran 1972).
19. For Stoic virtues, see Jedan 2009.
20. It should also be noted that critiques of extravagance are not specific to Seneca but go back to early Stoicism, as is borne out for instance by Plutarch's polemic against Chrysippus discussed above.

Chapter 5

1. Here and in the rest of the chapter, I use the term 'to express' in a sense analogous to the use in genetics, but with an ethical connotation: what is expressed is not just coded information but a natural orientation or inborn motivational disposition, which is not just transformed into biological structures and behaviours, but realized and achieved as the development of a potential, whose actualization is valuable or good for the being in question.
2. Properly documenting the description of Stoic physics and metaphysics given here would require an apparatus of footnotes far beyond the scope of this chapter. I therefore point the reader to the evidence provided in Wildberger 2006 and, concerning selected aspects, Salles 2009 and Wildberger 2008 and 2013.
3. Scholars disagree whether the Stoics regarded Matter as a hindrance to God's activity. The objection is that Matter is inert and thus cannot oppose God as a cause. However, it is possible to conceive of Matter's inertia not as an active cause but as something that just by existing and lacking the qualities of God makes a difference for the way in which God is active (Wildberger 2006, vol. 1: 51–5, 64–72, 276f.).

4. See, with regard to Plato, Chapter 3 by David Sedley in this volume.
5. For a seminal discussion of such questions see Mansfeld 1979.
6. The activity indicated in this source is specifically rational, goal directed action (*praxis*), not just any kind of motion or behavior. We may add, that God is the active principle, or 'that which makes/creates' (*to poioun*; see, e.g., Diog. Laert. 7.134).
7. This distinction is attributed to Chrysippus and is attested as such also in other sources (e.g. *SVF* 3.674).
8. For further evidence and discussion see Wildberger 2006, vol. 1: 209f.
9. Compare also his definition of the end in Diog. Laert. 7.88 = *SVF* 3.4. *Boulēsis* here is the divine equivalent to the eupathic impulse that consists in some sage's 'reaching' (*orexis*) directed at a good (for human *boulēsis*, see Graver 2007: 39–40).
10. See Chrysippus in Calcidius *in Tim.* 144 = *SVF* 2.933; compare also his definition of the end in Diog. Laert. 7.88 = *SVF* 3.4. *Boulēsis* here is the divine equivalent to the eupathic impulse that consists in some sage's 'reaching' (*orexis*) directed at a good (for human *boulēsis*, see Graver 2007: 39–40).
11. See also the nuanced discussion in Holland 1997: 153–7, and a more detailed account of Stoic conceptions of statehood and the teleological orientation of the Stoic state towards beneficence in Wildberger 2018: esp. 111–16.
12. Further references in Wildberger 2006, vol. 1:19f., vol. 2: 491f. n. 130.
13. See also Seneca, *On Benefits* 4.1–25 on the intrinsic value of beneficence for the person conferring a benefit and 6.21 on the essential beneficence of gods, like the Sun and the Moon, and the human sage: it is impossible for them not to want to be beneficent. The same point is made in shorter form at *Ep.* 95.49. In the same letter, Seneca presents the idea that humans are essentially sociable (*sociabiles*) because they are part of the same cosmic organism 'in which divine and human things are enclosed' (95.52).
14. In the quoted passage Cicero discusses phenomena on land and water. The motif continues through the whole account, and Balbus makes similar observations concerning the other two spheres, highlighting multitude and variety in the air and the aether (*Nat. D.* 2.101–15). Note also how the quoted passages point both to the whole (all = *omnium*; the overall view of the sea) and to the numerous individuals differentiated from each other by variation.
15. See *SVF* 3.74–8 for this definition and that of its opposite, what is bad or evil, as 'harm or nothing else than harm'.
16. Concerning *hupotelides* see Stobaeus, *Anthology* 2.2.3c; Wachsmuth 1884: 47, with Wildberger 2006, vol. 1: 207 and vol. 2: n. 986. This chapter is not the place to substantiate the analysis of goodness made here. Important contributions in that direction are, e.g., Bees 2004 (summarized in Bees 2011) and Gill 2006.
17. See above, p. 69, with relevant Latin text.
18. For the part-whole ontology, see Sextus Empiricus, *Against the Mathematicians* 11.24 = *SVF* 3.75.

19. In his critique, Plotinus ignores the Stoic ontology of parts and the complex functional reciprocity between the part and the whole (*Enn.* 1.6.1). He attempts a *reductio ad absurdum* by claiming that the parts would not be beautiful because they are not the beautiful whole and the whole would not be beautiful because it would be composed of non-beautiful parts. Roughly the inverse has been objected to the Stoics and the extensionalism of Deep Ecology, i.e. the idea that a proper ecological attitude implies self-extension in such a way, that the interests of the extended Self coincide with those of the ecological community as a whole, by Cheney 1989, whose reading elicited replies by Stephens 1994 and Velayos Castelo 1996 (on extensionalism see also, e.g., Plumwood 1993; Becker 2009 and Diehm 2010). In effect, this chapter is another reply since it argues that self-extension, identifying with the cosmos and God, implies precisely that one embraces and respects other individuals as no less essential to the beauty of the whole than the whole itself. The sage takes the 'cosmic viewpoint' (Williams 2012) not for the purpose of narcissistic self-aggrandizement but in order to become stronger and more capable to serve others.
20. See, e.g., Sandler 2007 and Chapter 4 by Christoph Jedan in this volume (pp. 49–61).
21. For Nussbaum and the Stoics, see Wildberger 2018: 216–28.

Chapter 6

1. Anthony Capon (University of Sydney), address to the Institute for Sustainable Places, Cardiff University, Cardiff, 16 April 2018.
2. Darwin wrote of Empedocles, 'we see here the principle of natural selection shadowed forth', on which see Burrow 1985: 53.
3. Teleology in the biological sciences is discussed at greater length in Nagel 1961.
4. Unless otherwise specified, all English translations of Greek and Roman works are taken from critical editions of the Loeb Classical Library (LCL).
5. Passmore does not supply a precise source for this quote from Basil the Great, the genuineness of which is disputed by some scholars.
6. There are also weaker forms of anthropocentrism, which either represent ethics as concerned with all objects of human interest or concern, or all human interaction with the world as anthropocentric. These more palatable varieties blunt the useful original distinction between ethics being limited to human interests and not being thus restricted. They also need to explain why things being objects of human interest or concern or of human interaction makes them morally significant, when all exactly similar objects which happen not to be objects of such interest or interaction lack moral significance altogether. In principle, these varieties of anthropocentrism, like the varieties depicted in the text, could be held, *mutatis mutandis*, either in metaphysical or in axiological form.
7. See further Attfield 2011.
8. On the phrase 'scale of nature' used by seventeenth-century writers, including Milton in *Paradise Lost*, see Lovejoy 1936: 89, 164 and 355 n.74.

9. Lovelock still remains a supporter of expanding nuclear energy generation (Lovelock 2006).
10. For other examples of New Testament narratives probably originating in parables, see Attfield 1983: 30.
11. For the Millennium Development Goals, see www.un.org/millenniumgoals/ (accessed 7 May 2018); for the Sustainable Development Goals, see www.un.org/sustainabledevelopment/sustainable-development-goals/ (accessed 8 May 2018).
12. This approach forms part of the narrative of religious organizations, e.g. the Board of Social Responsibility of the Church of England (Attfield 2015: 49–50), and secular ones, e.g. the Worldwide Fund for Nature (WWF) (Welchman 2012; Worrell and Appleby 2000).

Chapter 7

1. The gulf between the Old Testament and contemporary ecological ethics is stated most clearly by Rodd (2001: 237–49), though Wright offers a more optimistic assessment (Wright 2004). The ambivalent nature of the Hebrew Bible for the reconstruction of an ecological theology is recognized by many interpreters, though resolved in different ways. For a clear overview of different approaches, see Horrell (2010: 11–20, 117–21). Probably the most successful attempts to bridge this divide entail identifying key issues that enable dialogue between the text and modern environmental concerns, as for example through Horrell's identification of 'doctrinal lenses' (2010: 128), or Marlow's ethical model derived from her careful exegesis of passages from the pre-exilic prophets (2009).
2. According to this much-debated mode of interpretation of Gen. 1:26–28, humanity's relation to the rest of creation should be characterized as democratization of the kingly ideal, in which their rule is not one of domination but should foster the wellbeing of their subjects (e.g. Westermann 1988: 11). See Berry 2006 for an exploration of this topic, including evaluations of its validity and implications.
3. Highly anthropocentric understandings of the place of humanity and similarly elevated views of its status were long regnant in Old Testament scholarship and remain influential especially in conservative circles. This attitude, seen already in certain respects in patristic exegesis (Louth 2001: 28–9, 40 [Chrysostom], 44–5 [Gregory of Nazianzus]), was still dominant through to the early twentieth century. It is epitomized, for example, by Dillmann's 1892 Genesis commentary (for which the English translation appeared in 1897; see 78, 80, 85, 116). The sense that Eden is 'man's world …, which God … establishes *around man*' and that 'this forms the primary theme of the narrative' is still firmly articulated in the work of a scholar such as von Rad (1972: 76), and it is this tradition of interpretation that has lent credence to Lynn White's famous, but much-debated, placing of the 'historical roots of our ecologic crisis' in the anthropocentrism of medieval Christianity (White 1967). Though now largely rejected in scholarly circles, the endurance of a highly anthropocentric mindset rooted in this tradition of biblical interpretation is still reflected in the association of fundamentalism with low pro-environmental intentions (Bulbulia et al. 2016).

4. Job 38.16.
5. E.g. Pss. 46.3–4[2–3]; 65.7; 93.
6. E.g. Isa. 17.12–13, Jer. 6.23, 51.42, Zech. 10.11.
7. E.g. Isa. 51.15, Jer. 5.22, 31.35, Am. 5.8, 9.6, Nah. 1.4, Ps. 89.10[9], 104.6–9, 107.23–32, 135.6, Job 38.8–11, Prov. 8.29.
8. All biblical translations in the present chapter are from the NRSV, unless otherwise stated. Verse numbers in square brackets refer to versification in the Hebrew text where this differs from the English.
9. E.g. Gen. 15.18, Num. 34.3,5–7,11–12, Deut. 11.24, Josh. 1.4; cf. Ps. 72.8.
10. Exod. 14.10–29, Pss. 77.17–21[16–20], 78.13,53, 106.9–11, 136.13–15, Neh. 9.11.
11. See similarly Isa. 2.16, against the ships of Tarshish.
12. The significance of the alternation of the words 'ēl ('El', 'a god' or 'God') and 'ĕlōhîm ('God' or 'gods') in v.2 has been much debated, since this combination could be interpreted in a number of different ways. For a discussion of the issues involved, see van Dijk 1968: 92; Zimmerli 1983: 77–8.
13. Referred to by Philo of Byblos as 'Tyre, the holy island' (Eusebius of Caesarea, *Praeparatio evangelica* 1.10.32), the location of this city in the midst of the sea is alluded to in a variety of ancient Near Eastern and Classical sources (Katzenstein 1973: 9). It was later joined to the mainland via a causeway under Alexander the Great during the siege of 332 BCE.
14. The persistent allusions to the beauty of Tyre in Ezekiel 28 also hearken back to the thematic centring of the ship depiction around the idea of beauty. Note especially in this regard Ezek. 27.3–4,11e; the detailed description of her construction in 27.5–11 comprises an elaboration and substantiation of this framing motif of being made perfect in beauty, as indeed may also the depiction of the luxury products in which she traded in vv.12-25a.
15. If Boyle is correct, the 'heart of the sea' may also specifically be a place of divine judgment (2004: 17–27).
16. Cf. Ezek. 28.13,14,16. For mountains as cosmic centres, deeply bound up with Temple symbolism, see Clifford 1972, Keel 1997: 113–20, Wyatt 2001: §5. However, mountains in the sea may also be thought of as supporting the heavens: compare Ps. 18.8[7] with 2 Sam. 22.8 and cf. also Ps. 18.16[15]//2 Sam. 22.16, Prov. 8.27; see the further discussion in Keel 1997: 20–5, illuminating the notion of these mountains as islands and as providing points of connection with the worlds above and below.
17. The association of the 'heart of the seas' with the Pit in Ezek. 28.8 finds a further echo in Jon. 2.4[3]. The remaining three instances in which humans may be implicitly imagined as in or near the 'heart of the sea(s)' reinforce the sense that this is an extreme environment not intended for humanity: there is a reference to the congelation of the deeps in the heart of the seas at the Red Sea (Exod. 15.8), but not in this context to a crossing, though for the pursuing Egyptians descent into the sea leads to death (vv. 9–12); the way of a ship in the heart of the sea is described as a mystery 'too wonderful for me', something 'I do not understand' in Prov. 30.18–19; and drunkenness is described as like lying in the heart of

the sea in Prov. 23.34. Cf. *EA* 288.33, where to be 'situated like a ship in the midst of the sea' is to be isolated and surrounded by danger.

18. See Hummel's comment: '[bəlēb yamîm], "in the heart of the seas," must be a reference to Tyre's insular situation, its remoteness strengthening the picture of godlikeness beyond human reach' (Hummel 2007: 845). However, Tyre is not normally perceived as remote, but as a trading hub, the gateway to the peoples (Ezek. 26.2). Hence the remoteness is an expression of godlikeness rather than a coincidental aspect of reality supporting a theological view.
19. See Clifford 1972: 169–70, and Pope 1955: 98–9. Cf. Gilgamesh's quest to find Utnapishtim 'at the mouth of the rivers' in Gilgamesh IX–X: the aspects of inaccessibility to mortals, location across the sea, a garden of precious jewels, and guardian beasts all echo Edenic motifs. See Horowitz 1998: 96–106.
20. The close association of Eden and Temple symbolism is well established; see, e.g., Morales 2014.
21. The endurance of this motif in Temple and ecclesiastical flooring from ancient through to medieval times is traced by Fabio Berry (2007: 627–56).
22. Translation mine; the Hebrew reads *yhwh lammabbûl yāšāb*.
23. Cf. similarly Ps. 104.3 and perhaps also Ps. 104.13 and Am. 9.6.
24. See e.g. Psalms 29, 150.1. For the fusion of cultic and cosmological symbolism and thought, see, e.g., Morales 2014.
25. E.g. Pindar, *Isthmian Ode* 4.29–31, *Nemean Ode* 3.20–23, *Olympian Ode* 3.43–45; Euripides, *Hippolytus* 742–50.
26. Specific comparisons with aspects of Near Eastern thought are made in Beaulieu 2016: 9–10.
27. For wisdom as a divine quality, see Isa. 31.2, Jer. 10.7, Dan. 2.20, 5.11; cf. 2 Sam 14.20; and note especially Ezek. 28.2,6. Intriguingly, the beauty of Tyre is not called into question, yet although the noun *yip'āh*, 'splendor', is in the Hebrew Bible predicated only of the ruler of Tyre (Ezek. 28.7,17), the verbal counterpart is applied solely to YHWH, of his divine radiance or 'shining forth'. This aspect of the divine is well attested in Mesopotamia: see Oppenheim 1977: 98. Similarly, although the ship Tyre is 'given honour' in 27.10, in the only other context in which *hādār* is the indirect object of the verb *nātan*, 'give', namely in the Dead Sea Scroll 4Q416 2.3.10, the direct object is God. 'Perfect in beauty' echoes Jerusalem's unique status: Ezek. 16.14, Ps. 50.2, Lam. 2.15 (noting again the affinities between Temple, Eden and divine abode). This language not only provides links between the imagery of Ezek. 28.1–9 and vv.10–19, but wisdom and beauty are also the attributes manifested in the fruit of the forbidden tree, which was good to look at and to be desired to make one wise (Gen. 3.6).
28. They ultimately reappear on the foundations of 'the holy city Jerusalem' in Rev. 21.19–20, though they are now increased from nine stones to twelve. Further correlations may be made between the construction of the ship Tyre in Ezek. 27.3–11 (especially vv. 5–9) and that of Temple and Tabernacle (Geyer 2004: 51–5, and Lee 2016: 90–5). For Tyre's

command of the cedar forests as a divine quality (27.5), see Wilson 2013: 256. For a summary of the points of connection between Eden and Temple, see Wenham 1985: 19–25 and for both as the sacred mountain, see Levenson 1976: 7–36.
29. Cf. Isa. 27.1, Ps. 74:14, Job 3.8.
30. The Hebrew could either describe how God plays 'with him [Leviathan]' (thus e.g. Day 1985: 72–3; Kraus 1993: 303; cf. Job 40.29 [41.5]), or Leviathan's play 'in it [the sea]' (thus e.g. NRSV, JPSA, Seybold [1996: 407], Allen [1983: 25]).
31. The Hebrew *kullām*, 'all of them' in v.27 probably pertains to the antecedent *kullām* in v.24a, referring to 'your works, the things you have made', i.e. (or at least including) 'your creatures'(*qinyānekā*, v.24c). Otherwise, it would refer more immediately to the manifold and diverse life of the sea enumerated in vv.25–26.
32. On the form and nature of Leviathan, see Watson 2005: 333–5, 342–66.
33. Cf. the notion in the Ugaritic texts that Yam was the 'darling' one of El (*KTU* 1.3.35–36), or indeed that Mot (*KTU* 1.4.viii.24) and the monster Arsh (*KTU* 1.3.iii.43), were the 'darling' ones of the gods.
34. Or, perhaps, has no ruler (*mōšēl*).
35. Translation mine. The Hebrew reads *melek 'al-kol-bǝnê-šāḥaṣ*.
36. The relationship between the perspectives and traditions represented by these two passages is a fascinating issue that cannot be explored in the limited space available here. It is possible that the allusion to Leviathan being put on a leash as a girl's plaything in Job 40.29 [41.5] could parody the tradition represented by Ps. 104.26.
37. For a summary of the issues surrounding the very large numbers in Numbers, see Cole 2000: 78–82. This includes the possibility that *'elep* may not even denote a thousand, but something else, such as a 'clan' or a military troop or division, or that the numbers are symbolic, or purposefully hyperbolic, or represent some form of gematria, or perhaps relate to population numbers at a subsequent time. The interpretation of *'elep* as denoting a 'troop', as calculated by Colin J. Humphreys (1998: 196–213 and 2000: 323–8) results in a much more modest total of about 20,000 people, including women and children.
38. Translation mine. The Hebrew reads: *hăṣō'n ubāqār yiššāḥēṭ lāhem umāṣā' lāhem 'im 'ēt-kol-dǝge hayyām yē'āsēp lāhem umāṣā' lāhem*.
39. Cf. Jn 21.11, in which a miraculous catch amounts to 153 large fish.
40. Literally, 'gather'. Baruch A. Levine infers that the reference may therefore be to an ancient method of fishing, the gathering of fish on the shore (1993: 325). However, the thematically significant repetition of the verb *'sp*, 'gather', in the course of this pericope (vv. 16, 24, 30, 32 [twice]; cf. v.4) may have been the primary impulse informing the choice of language (see Cole 2000: 183).
41. Thus NRSV.
42. Cf. similarly NJB.
43. This is communicated especially clearly by Levine: 'Could flocks and herds be slaughtered for them in quantities sufficient for them? Were all the fish of the sea to be caught for them, would that meet their needs?' (1993: 313); and by the JPSA: 'Could all the fish of the sea be gathered for them to suffice them?'

44. Cf. also Jer. 16.16.
45. Translation mine; Hebrew *lə'ēt rā'āh*.
46. This last phrase (not bracketed in the NRSV) is not directly expressed in the Hebrew, but the sense is implicit in the context.
47. The passive form of *'āḥaz,* the niphal participle.
48. Although *ra'* is omitted from the Vulg. and Tg. after *məṣôdāh* ('net'), it has strong support from the MT and LXX, as well as a more plausible reason for omission (homoioteleuton) than for addition. *Ra'* is, of course, a repeated motif in Qoheleth (Ecclesiastes), occurring thirty two times.
49. This might be understood in a broader cultural context than Qoheleth itself: note the discussion of the net as an agent of death and tragedy in Yoder 2016: 121–4.
50. On the fishing equipment implied here, see Yoder 2016: 32–3, 37–9. For the net as representing divine retribution and drawing on a wider complex of ancient Near Eastern divine and royal symbolism, see *idem*, 44–52, 77–86.
51. Translation mine; for a discussion of the semantic issues attending this poetic passage, especially as regards verbal tense, see Andersen 2001: 170–4.
52. Note the association of *dāg*, 'fish' with the verb *dāgāh*, 'to increase, multiply', and indeed the motif of innumerability vis-à-vis aquatic life in Ps. 104.25, Gen. 1.20–22, Exod. 7.28[8.3].
53. Following the MT and versions; 1QpHab has *lmšl* in place of *lō'-mōšēl*.
54. In the Hebrew Bible, the verb *māšal* 'to rule' is employed of God's kingship (Ps. 66.7) and of human rule over fellow countrymen (Josh. 12.2,5, Judg. 8.22, 9.2, Isa. 16.1, 52.5, Jer. 22.30, 30.21, 51.46, Zech. 6.13; and, in an adopted country, Gen. 45.8,26). 'Rule' may also be exercised over other members of a household (Gen. 24.2), or by the 'great lights' over the day and night (Gen. 1.18), or even in 'governing' one's temper (Prov. 16.32). It should entail responsibilities towards the subjects and benefits for them in return, though of course there were bad rulers who failed in this (e.g. Prov. 28.15, 29.2,12).
55. Contrast the positive assessment of the capability of ants and locusts respectively, despite their lack of a ruler or king, in Prov. 6.6–8 and 30.27, and cf. Job 41.25[33], where, if this root (*DCH* I *mšl*) rather than the sense of 'likeness' (II *mšl*) is intended, the absence of a 'ruler' over Leviathan is a mark of its prestige.
56. The terminology of 'tenor' and 'vehicle' originated with Ivor Armstrong Richards (1936).
57. This term is highly problematic since, although somewhat ubiquitous in the discussion of Psalm 8, as of Gen. 1.26–28, it is not a fair representation of the Hebrew employed here, even whilst it eases the discomfort of talking about human 'rule' or of the disregarding of species barriers which 'leadership' implies. See Berry 2006, and Bauckham 2010: 2–12 and the further references cited there.
58. Besides the passages concerning fishing discussed here, those portraying judgment on humanity as impacting on the non-human creation, as in Hos. 4:1–3, should also be considered.
59. See especially Job 38.39–41, 39.1–8,26–30, Pss. 104.10–30, 145.8,15–16 (noting the wider context of vv.12–17, which seems to place provision for the creatures within the

perspective of God's kingdom (*malkût*) and dominion (*memšālāh*)), 147.8–9 and Wisd. 11.26. See also Ps. 36.7[6], 136.25, and note the contrast in Sir. 18.13 between human compassion for neighbours and God's for every living thing.

60. Hence Haak's contention (1992: 50–1) that Hab. 1.14 may constitute a reversal of Ps. 8.7–9 has some validity, though of course the primary concern is with humanity's relation to God, not to its fellow creatures.

Chapter 8

1. For a semantic illustration of *chthon,* the most reliable source remains *LSJ* s.v. For the problematic entry in Chantraine, see p. 104 and p. 167 n. 9. For a recent survey and analysis of the deities attested in Greek sources as 'of the earth' (*chthonioi*), see Parker 2011: 81–2 with further bibliography. In relation to the two aspects of the chthonic, Parker notes that 'the double aspect of earth, as home of the dead and the source of growth, is well established; so too that those two aspects to some extent blend into one another.' See also Burkert 1985: 199–203; Scullion 1994 *passim.*
2. For these festivals, see Parke 1977; Parker 2005. It goes without saying that, as this was not the only dimension of these festivals, these deities' identity was not exclusively chthonic. See Scullion 1994.
3. The scholarship on this long-standing polarity, originating in the eighteenth and nineteenth centuries, is enormous; it has recently been surveyed by Deacy 2015: 356–60. Among the most influential contributors to the debate in recent decades were Schlesier 1991/1992, who argued that the term 'chthonic/chthonian religion' should be discarded; and Scullion 1994, who defended the 'fundamental soundness of the distinction between Olympian and chthonian gods and rituals' (117).
4. For the unilateral semanticization of the chthonic as 'of the underworld' with negative connotations, see, e.g., Dowden 2007: 47; Felton 2007: 90; Iles Johnston 1999; Schlesier 1997. This emphasis is upheld in most readings of the *Oresteia,* e.g. Garvie 1986: xxxi–xxxiii, Sommerstein 2010: 193–203; Rosenmeyer 1982: 365–8. For the connection with agriculture, see, e.g. Deacy 2015. Even Parker reduces growth to an agricultural process when he talks about chthonic powers (Parker 2011: 81). However, the title and content of a chapter in his earlier *Polytheism and Society,* 'Gods at Work II: The growth of plants and men' (2005: 416–51), suggests a more wide-ranging understanding of the phenomenon of growth. Another balanced, but little known, analysis is Fischer 2010.
5. Again, the scholarship is enormous; but see especially Ekroth 2002 and Hägg and Alroth 2005.
6. Both in 2011 and in 2005: 424–5. Cf. Schleiser 1997, who claims that the concepts 'chthonic religion', 'chthonic cult' and 'chthonic sacrifice' are now rarely used.
7. Parker 2005: 424–5.
8. The spectrum of variation in an indicative sample from the scholarship of the last forty years is suggestive and not only due to the different interpretative angles: Hadzisteliou-

Price 1978 *passim*, Boedeker 1984: 10, 96, 123; Kearns 1989 *passim*; Lloyd-Jones 1990; Henrichs 1991; Loraux 1993: 61, 223, 227–30; Bremmer 1994: 15; Daraki 1999 *passim*; Ekroth 2002 *passim*; Zografou 2010 *passim*; Catenaccio 2011; Deacy 2015.

9. Chantraine 1968 s.v. χθών; Wilamowitz-Moellendorff 1931: 210–11. Segal 1998: 205 for a classic statement of the position.
10. The nourishing and generative aspects of the earth are evoked in many archaic and classical texts when *chthon* or *chthonios* are used: Hesiod, *Works and Days* 465–6; Aeschylus, *Eumenides* 902–6. *Chthon* can also designate the whole world: Aechylus, *Persians* 75–6.
11. Burkert 1985: 200–1.
12. There is also the larger question of how we categorize deities who sometimes appear associated with the chthonic, like Zeus, Athena and most other Olympian gods. The most balanced approach is Scullion 1994.
13. Boys-Stones and Haubold 2010; Hunter 2014; Stamatopoulou 2017; Van Noorden 2015. On Hesiod's influence on the Milesian philosophers see Stokes 1962 and 1963; Solmsen 1949 and now Gregory 2016: 11, 58, 201–2.
14. See especially Nelson 1998.
15. Cf. Clay 2003: 88–9.
16. The principal study remains Solmsen 1949.
17. Scullion 1994: 111.
18. For bibliography and surveys of influential approaches, see the relevant entries in Roisman's encyclopedia (2014). Among the most cited are Goldhill 2004; Lebeck 1971; Sommerstein 2010; Taplin 1977; Zeitlin 1978.
19. For example, Garvie 1986: xxvi–xli; Mitchell-Boyask 2009: 27–33; Schlesier 1991/1992: 46–7; Scullion 2005; Sommerstein 2010: 193–203; Winnington-Ingram 1983: 154–74. A recent survey of the role of the Erinyes in Greek drama is Sewell-Rutter 2007: 78–109.
20. See, for example, Garvie 1986: xxxi–xxxiii; Scullion 2005; Solmsen 1949: 187–90; Winnington-Ingram 1983: 154–74; contra Mitchell-Boyask 2009: 29; Sommerstein 2010: 199.
21. See Sommerstein 1989 on *Eum.* 235–98 and 1021–47; cf. Mitchell-Boyask 2009: 55. For Athena Polias' chthonic links, see Scullion 1994: 86, 89–90, 106–9.
22. The first sacrifice of the pregnant hare and its young, which brings about the 'second sacrifice' (*Ag.* 150) of Iphigeneia, also has chthonic characteristics (*Ag.* 134–55). See Scullion 1994: 96, 97–8, 111–12. Although it has been shown that chthonic and Olympian sacrifice rituals can share elements, the sacrifice of pregnant victims tends heavily to be associated with deities' connections to earth, growth, death, etc.
23. Scullion 2005; Winnington-Ingram 1983: 154–74.
24. Bacon 2001; cf; Easterling 2008 and the 2006 *Metis* volume edited by Labarrière, J. -L. et al. which focuses on the Erinyes as key in the *Oresteia*; see also B. Fowler 1991; Padel 1992: 168–92; Taplin 1977: 407; Winnigton-Ingram 1983: 154–74.
25. Bakola 2018.
26. Bakola 2014 and 2016.
27. Winnington-Ingram 1983: 165.

28. Some of the most thorough analyses include Bacon 2001; Easterling 2008; Mitchell-Boyask 2009: 87–96; Rehm 2002: 97–100; Sommerstein 1989: 239–86; Winnington-Ingram 1983: 164–74.
29. Henrichs 1991; Lloyd-Jones 1990.
30. Brown's (1984) theory is now generally accepted; e.g. Henrichs 1991; Lardinois 1992; Parker 2009: 150–1; Sommerstein 1989: 9–12. *Contra* Lloyd-Jones (1990).
31. Rosenmeyer 1982: 366: 'The exchange and reconciliation of the Furies are, structurally, an afterthought. They are very exposed, not part of the body of the drama, but a special epilogue, outdoing the rationalizations, the *aitia,* of Athena and Orestes with a super-*aition*.'
32. See above n. 25.
33. Fraenkel 1950: 92–4; Furley 1986 sees in 'Wrath' the family Curse, which is essentially identified with the Erinys. However, because *Ag.* 154–5 is more commonly understood in relation to Clytemnestra and her famous 'control of the house', the key characterization *oikonomos* has not been considered in relation to the Erinys: e.g. Raeburn and Thomas 2011: 83; Sommerstein 2008: 19. Nevertheless, it is beyond doubt that *Menis* (i.e. 'Fury') at *Ag.* 155 refers as much to the Erinys/Curse of the house as to Clytemnestra.
34. The Greek text and translations of Aeschylus follow Sommerstein 2008, unless otherwise indicated.
35. See esp. *Ag.* 376–84. As Lebeck 1971: 37 acutely observed, in this ode's complex array of images and reflections, Paris and Agamemnon mirror one another in terms of guilt, and Menelaus is also entangled in the same nexus of blame. For the semantics of *koros* in this ode, see Noussia-Fantuzzi 2015: 61–5, Anhalt 1993: 90–3, Balot 2001: 89–93, and Helm 1993. Theognis uses *koros* in the same sense in frr. 605–6 and 693–4; (cf. Helm 1993: 10–11; Nagy 1985: 60–1). Crucially, *koros* means insatiability in Aeschylus' *Persians* (cf. Noussia-Fantuzzi 2015: 62–5; Rosenbloom 2006: 108–9, 134), a play which develops the discourse on wealth, *dike* and the earth in a way strikingly similar to the *Oresteia* (see pp. 112–13).
36. This is an idea predominant in Solon, another poet whose influences on this ode, and the Aeschylean representation of wealth as a whole, are profound, and appropriated from Hesiod, cf. *Op.* 240; cf. Noussia-Fantuzzi 2015.
37. See Collard 2003: 138. There is a series of identifications which construct the image of the Erinys and her avenging action in this ode: Erinys ~ Wrath in *Ag.* 702–3; Helen ~ lion cub (as a primary reference) ~ priest of *Ate* ~ Erinys in *Ag.* 717–49; black *daimon* of *Ate* ~ Erinys in *Ag.* 766–72. It is not only Paris and the Trojan household in our text that are punished by the power represented by the Erinyes. In the case of Menelaus, the horrific storm that eliminates the entire fleet of the Atreid brothers and, as we are meant to understand, takes Menelaus' life also, is envisaged, in fact, as a victory-hymn of the Erinyes (*Ag.* 645).
38. See Bakola 2016; Goldhill 1986: 11; Jones 1962: 82–93; Taplin 1977: 313–4.
39. Cf. Goheen 1955: 115–26, Lebeck 1971: 85; Scodel 1996: 120.
40. The Greek text here follows Page 1972. The translation has been adapted from Collard 2003 and Sommerstein 2008.

41. The analysis above is a summary from Bakola 2016. For Clytemnestra as the mouthpiece of the Erinys, who lures the human agents into hubris against wealth, see also Bakola 2018: 178–80.
42. At least on an ostensive level; however, see below, n. 60.
43. The spectacular imagery of the 'yoking' of the earth (*Pers.* 66–80; 130–1; 181–96; 722; 746–8) powerfully captures the human attempt for absolute control of the environment (cf. *Pers.* 109–13). The landscape, made strikingly animate, has a central role in the relentless revenge of the earth: cf. *Pers.* 272–7; 302–10; 482–512; 595–7; 792–4, and so do the natural phenomena and elements (*Pers.* 495–505). The earth's reaction is foreshadowed in the queen's words and actions, especially her attempts to propitiate the chthonic powers: *Pers.* 164; 195–8; 219–23.
44. The play insists on the youth of the men lost: *Pers.* 59–62; cf. 12–3; 33–4. Contrast with the old age of the chorus, the men left behind (*Pers.* 5 and *passim*).
45. The idea that human lives are mothered by the earth, evoked, as so often in Aeschylus, in its cultic perception as *kourotrophos* (youth-nurturing), emerges prominently: *Pers.* 59–62; 511–12; 923–4. It does not have only cultic, mythical or ideological associations; rather it is integrated into the rich economic fabric of the play. For ge *kourotrophos* (youth-nurturing earth) in Attica and for other divine manifestations of the nursing power of the earth, see Hadzisteliou-Price 1978: 101–32; Parker 2005: 426–40.
46. It has rightly been noted that this association is reworked by Sophocles in *OT*: see Segal's insightful analysis of the 'earth' in *OT* (1998: 199–212).
47. The final scene crystallizes this identification, revealing the two brothers' ultimate motive for fighting each other. However, this might have been suggested explicitly since the outset of the play. At *Sept.* 73 some manuscripts transmit a reading which suggests that Eteocles prays to Zeus, earth, the gods of the *polis* and the Erinys to protect *his flowing wealth* alongside his house, his city and his land – but many editors since Dawe 1964: 180–1 have deleted this verse. However, Hutchinson 1985 ad loc. is more cautious. The variant merits further investigation.
48. Cf. also *Sept.* 710–11, 727–33, 788–90, 816–9, 906–9, 941–50. The continuation of the family line and property is, theoretically, a strong motive for both brothers. However, the play chooses to exploit the *contrast* between this pursuit (i.e. the acquisition and control of the royal *oikos*, the land of Thebes and its wealth) and the destruction of life (i.e. wealth) that results from it. See Seaford 2013: 168–72.
49. The Greeks did not have a word for 'natural environment', and although this concept is captured to a degree by terms like *ge, chthon, phusis* in different texts, none of these words should be expected to capture it in its entirety. Nevertheless, in Aeschylean drama at least, language, imagery and dramatic space contribute together to the construction of this entity. See Bakola 2014: 25–9.
50. Identifying a link between earth and cosmos in the *Seven Against Thebes* is less straightforward, because we essentially have only one third of the whole work whilst both

Persians and the *Oresteia* are arguably complete. See, however, Seaford's excellent analysis of the representation of the earth in relation to the play's cosmology in 2013: 168–77.
51. For early Greek philosophy and tragedy, see Lloyd-Jones 1971: 79–103. For Aeschylus' influence by the Presocratics, see, in particular, Seaford 2013. For Solon and Aeschylus, see Noussia-Fantuzzi 2015.
52. For the relationship between divine agency and agency of natural elements in archaic Greek thought, see Gregory 2016: 97–101.
53. The classic study is Peradotto 1964. See also Scott 1966 for wind and storm imagery and Heath 1999 for animal imagery.
54. See Bakola 2018: 171–4.
55. Among the passages that connect Erinyes and the element of wind (as wind, storm, air, or breath) are the following: *Ag.* 182–202, 214–15, 218–27, 376–82, 634–5, 985–97, 1078–83, 1215–19, 1233–6, 1309; *Cho.* 33–41, 201–3, 391–3, 819–25.
56. In *Seven Against Thebes*, the Curse/Erinys had also been portrayed like a strong wind driving Eteocles to his mad duel with his brother and their death: 'the *daimon* may perhaps, given time, change the wind of your spirit, and blow with a gentler breath; but at present it is still seething … ' (*Sept.* 705–8)
57. See Lebeck 1971: 50, and above, n. 37.
58. See Bakola 2018: 180–2.
59. A detailed analysis of the role of the Erinyes and their connection with, or manifestation through, elements of the natural world is the subject of a monograph (in progress) entitled *The Erinyes and the Wealth of the Earth: Cosmos, Nature and Resources in Aeschylean Theatre*.
60. Whether the representation of what happens to the Erinyes is unambiguously celebratory is a complex matter, which cannot be discussed within the confines of this chapter. The fact that the Erinyes are offered celebratory clothing that is dyed with wasteful murex purple dye (*Eum.* 1028) gives us food for thought that the ending is not unambiguously optimistic about the relationship between humans and earth.

Chapter 9

1. However, Leitenberg points out the difficulty of reaching agreement on this since no two authors use the same definitions, criteria or sources. In his view, these figures are grossly understated since they take no account of violent deaths due to genocide and forced enslavement or cultural reform, e.g. in China during the Cultural Revolution, where Leitenberg estimates that 46 million people died between 1950 and 1956 (Leitenberg 2006: 3–5). He therefore revises upwards the estimates to reach a sobering total of between 225.5 and 237.5 million lives lost in the twentieth century through violence (Leitenberg 2006: 14).
2. To what extent there is a causal relationship between environmental issues such as climate change and increased conflict is debated; see discussion in Theisen et al. 2013.

3. As Paul Arnould puts it: 'Quatre ans de guerre peuvent être mis en parallèle avec le temps géologique de la dernière période froide de l'ère quaternaire.' 'Four years of war can be set alongside the geological time-frame of the last cold period of the Quaternary Ice-Age.' (P. Arnould, cited in Puyo 2004: 578)
4. http://www.nytimes.com/1990/11/13/world/hold-your-breath-it-s-france-s-scariest-job.html; http://www.dailymail.co.uk/news/article-2497732/The-iron-harvest-Meet-soldiers-tasked-clearing-hundreds-tonnes-deadly-World-War-I-shells-mines-beneath-fields-Flanders.html (accessed 9 August 2017).
5. Unless otherwise stated, all Bible references are from the New Revised Standard Version (Oxford 1989).
6. The written prophets of the Hebrew Bible (*nevî'îm 'aḥaronîm*) comprise the three major prophets: Isaiah, Jeremiah and Ezekiel, plus the Book of the Twelve: Hosea, Joel, Amos, Obadiah, Jonah, Micah, Nahum, Habakkuk, Zephaniah, Haggai, Zechariah, Malachi (called the twelve minor prophets by the Christian church). Only Haggai and Malachi, which are clearly set in a post-exilic context, contain little by way of depiction of the threat of invasion or of warfare.
7. The term 'war poem' is coined by Kathleen O'Connor of the cluster of poems in the book of Jeremiah that are dominated by war imagery, especially Jer. 4:5–6:30 (O'Connor 1988: 48). It is used here of any poem from the prophetic corpus that depicts war and its consequences.
8. The detailed biblical accounts of the conquest of both the Northern Kingdom Israel (2 Kings 17) and the Southern Kingdom Judah (2 Kings 24–25) are corroborated (in broad and more measured terms) by extra-biblical evidence such as the annals of the Assyrian king Ashurbanipal (*ANET* 280–82), and the Babylonian account of the conquest of Jerusalem (*ANET* 563–64).
9. The name of God, YHWH, known as the tetragrammaton, is derived from the enigmatic conversation between God and Moses in Exodus 3. In Judaism it is regarded as sacred, and therefore not to be spoken or written. In modern English translations such as the NRSV it is written as LORD.
10. On the lament form in the Hebrew Bible see Ferris 1992, Mowinckel 1962.
11. In canonical order: Isa. 24.1–20, 33.7–89; Jer. 4.23–28, 12.1–4, 7–13, 23.9–12; Hos. 4.1–3; Joel 1.5–20; Amos 1.2.
12. There is considerable overlap between these warfare/judgement accounts and more general theophany texts announcing the coming of YHWH that also depict the impact of his presence on nature (e.g. Ps. 18.7, Ps. 29.3–9).
13. The interplay between darkness and the day of YHWH in these texts is reflected in other prophetic texts (e.g. Joel 2.2, 4.14–15; Zeph. 1.15), see discussion in Marlow 2009: 198–9.
14. See Childs' discussion of v. 23 where the Hebrew verb *rāš* 'to quake' is used in conjunction with the enemy from the North. He argues that in exilic and post-exilic literature this root 'appears to have developed into a technical term for the final shaking of the world at the return of chaos' (Childs 1959: 197).

15. The traditional paradigm in literary studies that the war poets are a defined, unified group expressing outrage at the war has been heavily revised in the past two decades, see Vandiver 2010: 1–9.
16. This, one of Sassoon's earlier poems, idealizes and even romanticizes war, in contrast to the bitter satirical poems of his later years (Copp 2001: 30).
17. For a detailed examination of the use of motifs, images, modes of expression and even direct quotes from the Classics in First World War poetry, see Vandiver 2010.
18. The Hebrew word 'ereṣ (hā-'āreṣ) has a broad range of meaning from the cosmic (earth) to geographical/political (land) to local (ground) (Marlow 2012).
19. Ps. 37.5 reads 'I have been young, and now I am old, yet I have not seen the righteous forsaken or their children begging bread.' Blunden's re-working of this is prescient in the light of the refusal among some Jewish Shoah (Holocaust) survivors to recite this text as part of their grace after meals (Donin 1980: 284–304).

Chapter 10

1. LeVasseur and Peterson 2017 provides book-length reflection on White's influence, fifty years after publication; see in particular within this volume Whitney on White's cross-disciplinary reach. Cf. Jedan 2017, who likens the impact of White's article to a reformation. See also Jenkins 2009: 290–5 on White's influence on environmental ethics; Passmore 1974: 4–5 and Attfield 1983: 20–4 on environmental philosophy; Rigby 2002: 154–6 on ecocriticism.
2. For criticisms specifically of the *quality* of White's scholarship, see Attfield 1983: 23–4; Attfield 1994: 22; Barr 1972: 18; Hiebert 1996: 14; Marlow 2009: 14.
3. As Jenkins puts it, 'nearly every book on the relation of Christianity to its environment refers to White's thesis, and most introduce their argument as a definite response to it' (2009: 286). For examples of White being given the 'first word' in such scholarship, see Barr 1972: 15–16; Collins 1995: 87–8; Conradie 2010: 25; Hiebert 1996: 13–14; Horrell et al. 2010: 1–3; Marlow 2009: 11; Northcott 1996: 40.
4. Theological responses have been varied and creative, using trinitarian thinking to construct an understanding of God's involvement in the natural world (Moltmann 1985), developing eco-feminist ideas from a Christian perspective (e.g. Jantzen 1984), or calling for a radical shift in Christian cosmology with reliance on ideas central to deep ecology (Berry 1998). Some biblical approaches have homed in on 'stewardship' in Gen. 2:15 (e.g. Attfield 1983: 34–50), as well as rethinking 'dominion' (see Barr 1972: 19–23 and Bauckham 2010: 16–20). Some investigate texts other than Genesis, prompting both 'readings of recovery' and 'readings of resistance' (on which see Horrell 2010: 11–47): see Bauckham 2010: 64–82 on psalms; Marlow 2009 on prophets; Horrell 2010 on Paul.
5. As Marlow 2009: 14 argues, White's article 'lacks credibility from a biblical studies perspective', but is to be valued for the productive questions it raises. Cf. Habel 2000: 29–30 and Jenkins 2009: 283–4.

6. Whilst White refers loosely to 'antiquity', context strongly suggests he means the cultures of ancient Greece and Rome, whose mythology he referred to a few sentences previously.
7. For reasons of space, discussion will also be restricted to Anglophone scholarship.
8. See Hunt 2016: 29–71 for a detailed account, with a Roman and arboreal focus.
9. For an introduction to comparativist obsession with animism, see Lessa and Vogt 1972: 9–19; Sharpe 1975: 53–8 and 65–71; Stocking 1987: 188–97.
10. Scheid 1987: 303–5.
11. See Rose 1948: 9–11 for a classic (if late) statement of this stance, and Hunt 2016: 58–9 on the aligning of *mana* and *numen*.
12. Habinek 1998: 15–33 discusses Hellenic bias among nineteenth-century academics, which had a marked impact on the later study of Greek and Roman religion.
13. On the fall from grace of comparativism and animism see e.g. Gellner 1999: 19–20; Harris 1968: 258–9; Hunt 2016: 62–6; Lessa and Vogt 1972: 4–5; Whaling 1984: 177–8. Sourvinou-Inwood 1990 established the term *polis* religion.
14. See Hunt 2016: 62–71, again with an arboreal focus.
15. Hunt 2016 hoped to kickstart further in-depth work in this neglected area.
16. Hughes 2014 (revised version of book first published in 1994); Thommen 2012 (original language publication 2009); Vögler 1997; Weeber 1990. Glacken 1967: 147–9, summing up his exploration of the relationship between nature and culture in antiquity, concludes that Greek and Roman thinkers were not well attuned to 'the role of man in changing the environment' (148). Sallares 1991 takes a technological approach to the relationship between the Greeks and their environment, and is uninterested in worldviews. Interest in human–environment relationships within classical studies is, however, picking up momentum, as seen by the ICS Autumn 2016 seminar devoted to 'Ecology and Ancient History', or Bakola's forthcoming monograph on Aeschylean tragedy and its preoccupation with human relationships with the earth.
17. Hughes 2014: 6; Thommen 2012: 76.
18. Cf. Hughes 2014: 230.
19. Hughes 2014: 184 and 195; Thommen 2012: 42.
20. Turcan is extrapolating from three passages of Servius, Virgil and Tibullus, whilst Cole footnotes only Plato's *Phaedrus*. See also Turcan 2000: 42–3; Cole 2004: 30. Cole 2004: 64 notes that some places were 'so charged with divine power that they could not be entered at all'.
21. Cf. Cole 2004: 29.
22. I have found no evidence of 'pagan animism' being used in such a way before White.
23. What they would not find is engagement with White. Despite his bold claims about Greeks and Romans, ironically White is almost unheard of within classics and ancient history.
24. This section builds on the major arguments of Hunt 2016, to which the reader is referred for the full argument.
25. Within environmental philosophy and ethics significant academic interest has developed in the 'good' and the 'rights' of trees, e.g. Attfield 1994: 153–71; Stone 1972.

26. Tylor 1871: 196, the father-figure of animism, championed trees for their excellent 'illustrations of man's primitive animistic theory of nature'. See Hunt 2016: 50–4 on the privileged position that Greek and especially Roman examples held in comparativist scholarship on tree worship. Within the limited space devoted to the environment in contemporary portraits of Roman religion, trees are proportionally well treated, e.g. Rives 2007: 89–92; trees similarly star in Dillon 1997 on the 'ecology of the Greek sanctuary'; trees also attract more stand-alone studies than other features of the environment, e.g. Birge 1994; Henrichs 1979.
27. E.g. Dowden 2000: 76–7; Lane Fox 1986: 44.
28. For full discussion see Hunt 2016: 190–3.
29. On this passage's influence see Hunt 2016: 122–4 and 191–2.
30. Hunt 2016: 191–3.
31. For which see Hunt 2016: 193–6.
32. Hunt 2016: 177–84 on changing understandings of *numen* within scholarship on Roman religion, plus my argument that attempting to pin down *numen*'s meaning is inadvisable.
33. On which see Hunt 2016: 184–5, esp. n. 31.
34. Hunt 2016: 184–8 for detailed readings of the Ovid and Pliny passages.
35. See e.g. Henrichs 1979: 86; Thomas 1988: 263.
36. Discussed at Hunt 2016: 127–8.
37. Hunt 2016: 121–33.
38. Hunt 2016: 131–2 on the Lydian inscriptions; Hunt 2016: 137–52 on the Arvals' pruning.
39. Hunt 2016: 98–120 on the Ruminal fig.
40. Hunt 2016: 199–223.
41. Hunt 2016: 243–4 and 264–5.
42. Cf. Attfield 1983: 21 and 25.
43. See Marlow 2009: 52–70 for detailed discussion of this nature-history dichotomy in Christian thought. Hiebert 1996:1–22 works to prove that this distinction is neither watertight nor biblical.
44. Marlow 2009: 45–50.
45. Hodgson 1987: 238 and viii. See also Marlow 2009: 58–9 on Hegel's influence.
46. Marlow 2009: 67–8 on Wright. Wright talks interchangeably of 'pagan religions' (1952: 24) and 'ancient polytheists' (1952: 39), whereas some Old Testament scholars would privilege the term polytheist. But since 'pagan', as used by White and in classical scholarship, is a blanket term for polytheists (in opposition to Jewish and Christian monotheists), White's contrast of Christian vs pagan does map onto that of Israelite vs polytheist.
47. Passmore 1974: 10 for a brief statement of the position; Hiebert 1996: 3–22 for in-depth discussion.
48. Barr 1972: 15–16 and Hiebert 1996: 13 comment on this shift. See Jedan 2017: 3 on the ecological awakening in the years running up to White's lecture.
49. Bauckham 2010: 76–80.
50. Northcott 1996: 84–5 also sees the Middle Ages as a crucial turning point in Christian thinking about the environment, as of course did White.

51. Cf. Conradie 2010: 306 who suggests Christians explore common ground with 'indigenous or pre-Christian spiritualities' to recover a sense of nature's sacrality.
52. 'If new science brings God back into the picture, it is not the traditional God of the churches' (Schwarz 1991: 33).
53. Bauckham 2010: 79. Balabanski 2010 provides an interesting point of comparison. In a close reading of Colossians, she argues that Stoic elements of the pagan culture in which the church at Colossus grew up urged the Colossian Christians to a pantheistic but nevertheless Christ-focused view of the natural world ('in him all things hold together'; Col. 1.17). She further argues that this kind of view – terming it a 'permeation cosmology' – has real ecological potential for modern Christian thinkers in the value it places on the interconnectedness of creation. Yet, ecologically beneficial as it may be, Balabanksi still feels the need to stress that this is a Christological view, *despite* pantheistic *elements*. There is a tangible concern about appearing to promote a worldview that may have been overly influenced by surrounding pagan culture.
54. McFague 1993: 13–14.
55. Cf. the similar arguments of Jantzen 1984.
56. Like Moltmann, McFague uses the terms 'pantheism' and 'animism' in overlapping ways. Panentheism is also central to the ecotheology of Fox 1983.
57. McFague 1993: 143–50 sets out her understanding of this divine spirit.
58. See Harvey 1997: 167 and 170; York 2003: 13 and 62.
59. Cf. Harvey 2009: 406.
60. Harvey 2013: 1.
61. Harvey 1997: 175–6 and 2009: 393–4 notes how neopagans may use the terms 'animism', 'pantheism' and 'panentheism' in overlapping ways, with interesting parallels with the overlap of 'animism' and 'pantheism' in the work of Moltmann and McFague.
62. Cf. Hogan 2013: 19 and 26; York 2016: 329.
63. Cf. York 2003: 22.
64. Of course, White's dichotomy is by no means the only influence here, as I have shown. Perhaps the undermining of pagan animism from within classical scholarship might prompt experts in the Ancient Near East to join the debate, and to respond to understanding within biblical studies and Christian theology of ancient Near Eastern 'nature religions', held up as a foil to Israelite beliefs.
65. Some communication is clearly underway, but frustratingly the neopagan Taylor does not expand on his claim that 'contemporary animisms' are emerging from Christianity (https://animistjottings.wordpress.com/postmodern-animism-2/), nor the Spiral Goddess website on the claim that 'many environmentally conscious Christians today share the belief with Pagans that all forms of life have a soul' (http://spiralgoddess.com/Pagan101.html).

REFERENCES

Introduction

Bhagwat, S. A., N. Dudley, and S. R. Harrop (2011), 'Religious Following in Biodiversity Hotspots: Challenges and Opportunities for Conservation and Development', *Conservation Letters*, 4 (3): 234–40.
Branch, M. P., et al., eds. (1998), *Reading the Earth: New Directions in the Study of Literature and Environment*, Moscow, ID: University of Idaho Press.
Callicott, J. B. (1989), *In Defense of the Land Ethic*, Albany, NY: State University of New York Press.
Cittadino, E. (2006), 'Ecology and American Social Thought', in D. Lodge and C. Hamlin (eds.), *Religion and the New Ecology: Environmental Responsibility in a World in Flux*, 73–115, Notre Dame, IN: University of Notre Dame Press.
Crutzen, P. J. (2002), 'Geology of Mankind', *Nature*, 415 (6867): 23.
Dell, K. J., and W. Kynes, eds. (2012), *Reading Job Intertextually*, London: T & T Clark.
Francis (2015), *Laudato Sí: Encyclical Letter of the Holy Father Francis on Care for Our Common Home*, Vatican City: Libreria Editrice Vaticana.
Gottlieb, R. S. (2006), *The Oxford Handbook of Religion and Ecology*, Oxford: Oxford University Press.
Habel, N., ed. (2000), *Readings from the Perspective of the Earth*, Sheffield: Sheffield Academic Press.
Hailwood, S. (2004), *How to be a Green Liberal: Nature, Value and Liberal Philosophy*, Chesham, UK: Acumen.
Hinds, S. (1998), *Allusion and Intertext: Dynamics of Appropriation in Roman Poetry*, Cambridge: Cambridge University Press.
Horrell, D. G., et al. (2010), *Ecological Hermeneutics: Biblical, Historical and Theological Perspectives*, London: T & T Clark.
Jessop, R. (2012), 'Coinage of the Term Environment: A Word Without Authority and Carlyle's Displacement of the Mechanical Metaphor', *Literature Compass*, 9 (11): 708–20.
Kunkel, B. (2017), 'Capitalocene', *London Review of Books*, 39 (5): 22–28.
Lane, M. S. (2011), *Eco-Republic: Ancient Thinking for a Green Age*, Witney: Peter Lang.
Leopold, A. (1987), *A Sand County Almanac and Sketches Here and There*, Oxford/New York: Oxford University Press.
Marlow, H. (2009), *Biblical Prophets and Contemporary Environmental Ethics: Re-reading Amos, Hosea and First Isaiah*, Oxford: Oxford University Press.
Purdy, J. (2015), *After Nature: A Politics for the Anthropocene*, Cambridge, MA: Harvard University Press.
Rodd, C. S. (2001), *Glimpses of a Strange Land: Studies in Old Testament Ethics*, Edinburgh: T & T Clark.
Rolston III, H. (1995), 'Does Aesthetic Appreciation of Landscapes Need to be Science-Based?', *British Journal of Aesthetics*, 35 (4): 374–86.
Speth, J. G. (2008), *The Bridge at the End of the World: Capitalism, the Environment, and Crossing from Crisis to Sustainability*, New Haven, CT and London: Yale University Press.

References

Stoll, M. (2006), 'Creating Ecology: Protestants and the Moral Community of Creation', in D. Lodge and C. Hamlin (eds.), *Religion and the New Ecology: Environmental Responsibility in a World in Flux*, 53–72, Notre Dame, IN: University of Notre Dame Press.

Strong, M. (1993), *The Fourth Kew Environmental Lecture, Beyond Rio: New World Order, or Lost Opportunity?*, London: Royal Botanic Gardens Kew.

Thommen, L. (2012), *An Environmental History of Ancient Greece and Rome*, trans. P. Hill, Cambridge: Cambridge University Press.

White, L, Jr. (1967), 'The Historical Roots of Our Ecologic Crisis', *Science*, 155 (3767): 1203–07.

Williams, M. et al. (2011), 'The Anthropocene: A New Epoch of Geological Time?', *Philosophical Transactions of the Royal Society A*, Special issue, 369 (1938): 835–1011.

Wilson, K. (2008), *Jesus Brand Spirituality*, Nashville, TN: Thomas Nelson.

Wright, C. J. H. (2004), *Old Testament Ethics for the People of God*, Leicester: InterVarsity Press.

Chapter 1

Árnason, J. P., S. N. Eisenstadt, and B. Wittrock, eds. (2005), *Axial Civilizations and World History*, Leiden and Boston, MA: Brill.

Baldry, H. C. (1952), 'Who Invented the Golden Age?' *The Classical Quarterly*, 2 (1–2): 83–92.

Bellah, R. N. (2005), 'What Is Axial about the Axial Age?' *Archives Européennes de Sociologie/European Journal of Sociology/Europäisches Archiv Für Soziologie*, 46 (1): 69–89.

Boman, T. (1960), *Hebrew Thought Compared with Greek*, Philadelphia, PA: Westminster Press.

Brand, S. (1968), *The Whole Earth Catalog*, Self-published; electronic edition http://www.wholeearth.com/members/index.php.

Chakrabarty, D. (2009), 'The Climate of History: Four Theses', *Critical Inquiry*, 35 (2): 197–222.

Cooper, J. M., ed. (1997), *Plato: Complete Works*, Indianapolis, IN: Hackett.

Crutzen, P. J. and E. F. Stoermer (2000), 'The "Anthropocene"', *International Geosphere Biosphere Program Newsletter*, 41: 17–18.

Eisenstadt, S. N., ed. (1986), *The Origins and Diversity of Axial Age Civilizations*, Albany, NY: State University of New York Press.

El Murr, D. (2009), 'Hesiod, Plato, and the Golden Age: Hesiodic Motifs in the Myth of the *Politicus*', in G. R. Boys Stones and J. H. Haubold (eds.), *Plato and Hesiod*, 276–97, Oxford: Oxford University Press.

Fretheim, T. E. (2012), 'Genesis and Ecology', in C. A. Evans, J. N. Lohr, and D. L. Petersen (eds.), *The Book of Genesis: Composition, Reception, and Interpretation*, 683–706, Leiden and Boston, MA: Brill.

Glacken, C. J. (1967), *Traces on the Rhodian Shore: Nature and Culture in Western Thought from Ancient Times to the End of the Eighteenth Century*, Berkeley, CA: University of California Press.

Goldin, O. (1997), 'The Ecology of *Critias* and Platonic Metaphysics', in L. Westra and T. M. Robinson (eds.), *The Greeks and the Environment*, 73–80, Lanham, MD, New York, Boulder, and Oxford: Rowman and Littlefield.

Goldschmidt, E. E. (2013), 'The Evolution of Fruit Tree Productivity: A Review', *Economic Botany*, 67 (1): 51–62.

Haeckel, E. J. (1866), *Generelle Morphologie der Organismen*, Berlin: n.p.

Havrelock, R. (2017), 'Parashat Va-'era', in D. Lipton (ed.), *From Forbidden Fruit to Milk and Honey: A Commentary on Food in the Torah*, 99–101, Jerusalem and New York: Urim Publications.

Hillel, D. (2007), *The Natural History of the Bible: An Environmental Exploration of the Hebrew Scriptures*, New York: Columbia University Press.

References

Hughes, J. D. (1994), *Pan's Travail: Environmental Problems of the Ancient Greeks and Romans*, Baltimore, MD: Johns Hopkins University Press.
Kasimis, D. (2013), 'The Tragedy of Blood-Based Membership: Secrecy and the Politics of Immigration in Euripides' *Ion*', *Political Theory*, 20: 1–26.
Kolbert, E. (2014), *The Sixth Extinction: An Unnatural History*, London: Bloomsbury.
Lane, M. (2009), 'Thoreau and Rousseau: Nature as Utopia', in J. Turner (ed.), *A Political Companion to Henry David Thoreau*, 341–71, Lexington, KY: University Press of Kentucky.
Lane, M. (2011/2012), *Eco-Republic*, London: Peter Lang [2011]; Princeton, NJ: Princeton University Press [2012].
Lane, M. (2016), 'Political Theory on Climate Change', *Annual Review of Political Science*, 19: 107–23.
Lane, M. (2017), 'Parashat Bere'shit', in D. Lipton (ed.), *From Forbidden Fruit to Milk and Honey: A Commentary on Food in the Torah*, 23–24, Jerusalem and New York: Urim Publications.
Lipton, D. (2017), 'Va'era', in D. Lipton (ed.), *From Forbidden Fruit to Milk and Honey: A Commentary on Food in the Torah*, 102–06, Jerusalem and New York: Urim Publications.
Manning, E. N. (2010), *Hesiod and the Hebrews: The Greeks, the Hebrews, and the Western View of Nature*, Saarbrücken: VDM Verlag.
Nagourney, A., and H. Fountain (2017), 'Oroville Is a Warning for California Dams, as Climate Change Adds Stress', *The New York Times*, 14 February 2017. https://www.nytimes.com/2017/02/14/us/oroville-dam-climate-change-california.html
Passmore, J. A. (1974), *Man's Responsibility for Nature; Ecological Problems and Western Traditions*, London: Duckworth.
Rabbinical Assembly (2001), *Etz Hayim: Torah and Commentary*, New York: The Rabbinical Assembly.
Rackham, H. (1942), *Pliny, Natural History: Volume II*. Cambridge, MA: Harvard University Press.
Schlegel, C. M., and H. Weinfeld, trans. (2006), *Hesiod: Theogony* and *Works and Days*, Ann Arbor, MI: University of Michigan Press.
Schofield, M. (2006), *Plato. Political Philosophy*, Oxford: Oxford University Press.
Shaw, B. D. (1981), 'Climate, Environment, and History: The Case of Roman North Africa', in T. M. L. Wigley, M. J. Ingram, and G. Farmer (eds.), *Climate and History: Studies in Past Climates and Their Impact on Man*, 379–403, Cambridge: Cambridge University Press.
Thommen, L. (2012), *An Environmental History of Ancient Greece and Rome*, trans. P. Hill, rev. English edition, Cambridge and New York: Cambridge University Press.
Thoreau, H. D. (2000), *Wild Fruits*, ed. B. P. Dean, New York: W. W. Norton.
Tress, D. G. (1997), 'The Philosophical Genesis of Ecology and Environmentalism', in L. Westra and T. M. Robinson (eds.), *The Greeks and the Environment*, 33–42, Lanham, MD, New York, Boulder, and Oxford: Rowman and Littlefield.
West, M. L., ed. (1978), *Hesiod, Works and Days*, Oxford: Clarendon Press.
West, M. L., ed. (1998), *Aeschylus, Tragoediae*, Stuttgart and Leipzig: B. G. Teubner.
White, L., Jr. (1974), 'The Historical Roots of Our Ecologic Crisis', in D. Spring and E. Spring (eds.), *Ecology and Religion in History*, 15–31, New York: Harper & Row.

Chapter 2

Alexandre, C. (1856), *Excursus ad Sibyllina*, Paris: F. Didot.
Austin, R. (1927), 'Virgil and the Sibyl', *Classical Quarterly*, 21: 100–05.
Buitenwerf, R. (2003), *Book III of the Sibylline Oracles and its Social Setting*, Leiden: Brill.

References

Collins, J. J. (1983), 'Sibylline Oracles: A New Translation and Introduction', in J. H. Charlesworth (ed.), *The Old Testament Pseudepigrapha*, 1: Apocalyptic Literature and Testaments, 317–472, London: Darton, Longman and Todd.

Collins, J. J. (1997), *Seers, Sibyls and Sages in Hellenistic Judaism*, Leiden: Brill.

Erler, M. (1987), 'Das Recht (ΔIKH) als Segensbringerin für die Polis. Die Wandlung eines Motivs von Hesiod zu Kallimachos', *Studi Italiani di Filologia Classica*, 80: 5–36.

Fretheim, T. E. (2005), *God and World in the Old Testament: A Relational Theology of Creation*, Nashville: Abingdon Press.

Gauger, J. D. ([1998] 2011), *Sibyllinische Weissagungen*, Berlin: De Gruyter.

Glacken, C. J. (1967), *Traces on the Rhodian Shore: Nature and Culture in Western Thought from Ancient Times to the End of the Eighteenth Century*, Berkeley and Los Angeles, CA: University of California Press.

Kerridge, R. (2014), 'Ecocritical Approaches to Literary Form and Genre: Urgency, Depth, Provisionality, Temporality', in G. Garrard (ed.), *The Oxford Handbook of Ecocriticism*, 361–76, New York: Oxford University Press.

Lee, G. (1984), *Virgil: The Eclogues*, London: Penguin Classics.

Lightfoot, J. L. (2007), *The Sibylline Oracles. With Introduction, Translation, & Commentary on the First and Second Books*, Oxford: Oxford University Press.

Manietti, G. (1997), 'The Language of the Sibyls', *Euphrosyne*, 25: 237–50.

Marlow, H. (2009), *Biblical Prophets and Contemporary Environmental Ethics*, Oxford: Oxford University Press.

Marlow, H., K. Pollmann, and H. Van Noorden, eds. (forthcoming), *Eschatology in Antiquity*, London: Routledge.

Martindale, C. A. (1997), 'Green Politics: The *Eclogues*', in *The Cambridge Companion to Virgil*, 107–24, Cambridge: Cambridge University Press.

Nash, R. F. ([1967] 2001), *Wilderness and the American Mind*, New Haven, CT: Yale University Press.

Neutel, K. B. (2015), *A Cosmopolitan Ideal: Paul's Declaration 'Neither Jew Nor Greek, Neither Slave Nor Free, Nor Male and Female' in the Context of First-Century Thought*, London and New York: Bloomsbury.

Perkell, C. G. (2002), 'The Golden Age and Its Contradictions in the Poetry of Vergil', *Vergilius*, 48: 3–39.

Potter, D. (1994), *Prophets and Emperors: Human and Divine Authority from Augustus to Theodosius*, Cambridge, MA: Harvard University Press.

Ryken, L., J. C. Wilhoit, and T. Longman III, eds. (1998), *Dictionary of Biblical Imagery*, Leicester: InterVarsity Press.

Sandy, D. B. (2002), *Plowshares and Pruning Hooks. Rethinking the Language of Biblical Prophecy and Apocalyptic*, Leicester: InterVarsity Press.

Satterfield, S. (2016), 'The Prodigies of 17 B.C.E. and the *Ludi Saeculares*', *Transactions of the American Philological Association*, 146: 325–48.

Saunders, T. (2008), *Bucolic Ecology: Virgil's Eclogues and the Environmental Literary Tradition*, London: Duckworth.

Chapter 3

Diels, H., revised by W. Kranz (1952 and later editions; original edition 1903), *Die Fragmente der Vorsokratiker*, Berlin: Weidmann.

Gregory, A. (2007), *Ancient Greek Cosmogony*, London: Bristol Classical Press.

References

Lovelock, J. E. (1979), *Gaia. A New Look at Life on Earth*, Oxford: Oxford University Press.
Sedley, D. (2007), *Creationism and Its Critics in Antiquity*, Berkeley, CA and London: University of California Press.
Sedley, D. (2016), 'Empedoclean Superorganisms', *Rhizomata*, 4: 111–25.
Spencer, H. (1876), *The Principles of Sociology*, vol. 1, New York: D. Appleton & Company.
Wheeler, W. M. (1911), 'The Ant Colony as an Organism', *Journal of Morphology*, 22: 307–25.

Chapter 4

Algra, K. (2003), 'Stoic Theology', in B. Inwood (ed.), *The Cambridge Companion to the Stoics*, 153–78, Cambridge: Cambridge University Press.
Attfield, R. (2014), *Environmental Ethics: An Overview for the Twenty-First Century*, Cambridge: Polity.
Castelo, C. V. (1996), 'Reflections on Stoic Logocentrism', *Environmental Ethics*, 18 (3): 291–96.
Cheney, J. (1989), 'The Neo-Stoicism of Radical Environmentalism', *Environmental Ethics*, 11 (4): 293–325.
Corcoran, T. H., trans. (1972), *Seneca, Naturales Quaestiones. Books IV – VII*, Cambridge, MA: Harvard University Press.
Dragona-Monachou, M. (2010), 'Η φύσις των όλων και ο άνθρωπος στο Στωικισμό του Μάρκου Αυρήλιου από τη σκοπιά της οικο-ηθικής', unpublished draft.
Dragona-Monachou, M. (2015), 'Reason and Right Reason in Stoic Ethics', *Philosophical Inquiry*, 39 (1): 189–206.
Gill, C. (2016), 'Stoicism and the Environment', *Stoicism Today*, 6 February 2016, blog.
Holland, A. J. (1997), 'Fortitude and Tragedy: The Prospects for a Stoic Environmentalism', in L. Westra and T. M. Robinson (eds.), *The Greeks and the Environment*, 151–66, Lanham, MD: Rowman & Littlefield.
Jedan, C. (2009), *Stoic Virtues: Chrysippus and the Religious Character of Stoic Ethics*, London and New York: Bloomsbury.
Jedan, C. (2010), 'Göttliches und menschliches Handeln in der frühen Stoa', in R. Hofmeister Pich and J. Müller (eds.), *Wille und Handlung in der Philosophie der Kaiserzeit und Spätantike*, 25–44, Berlin and New York: De Gruyter Mouton.
Jedan, C. (2017), 'A Different Kind of Reformation: Revisiting the Lynn White Thesis', *NTT: Journal for Theology and the Study of Religion*, 71 (3): 277–86.
Jonas, H. (1979), *Das Prinzip Verantwortung: Versuch einer Ethik für die technologische Zivilisation*, Frankfurt am Main: Insel.
Lane, M. (2011), *Eco-Republic: Ancient Thinking for a Green Age*, Oxford: Peter Lang.
LeVasseur, T., and A. Petersen, eds. (2017), *Religion and Ecological Crisis: The 'Lynn White Thesis' at Fifty*, New York: Routledge.
Naess, A. (2003), 'The Deep Ecological Movement: Some Philosophical Aspects', in A. Light and H. Rolston III (eds.), *Environmental Ethics: An Anthology*, 262–74, Oxford: Blackwell, 2003.
Passmore, J. A. (1974), *Man's Responsibility for Nature: Ecological Problems and the Western Traditions*, New York: Scribner.
Plumwood, V. (1993), *Feminism and the Mastery of Nature*, London: Routledge.
Protopapadakis, E. D. (2012), 'The Stoic Notion of Cosmic Sympathy in Contemporary Environmental Ethics', in R. Vasić, D. Stefanović, K. M. Gadanski, and I. Gadansk (eds.), *Antiquity, Modern World and Reception of Ancient Culture*, 290–305, Belgrade: The Serbian Society for Ancient Studies.
Rackham, H., trans. (1933), *Cicero, On the Nature of the Gods. Academics*, Cambridge, MA: Harvard University Press.

Sandler, R. D., and P. Cafaro, eds. (2005), *Environmental Virtue Ethics*, Lanham, MD: Rowman & Littlefield.
Sedley, D. N. (2007), *Creationism and Its Critics in Antiquity*, Berkeley, Los Angeles and London: University of California Press.
Sessions, G. (1977), 'Spinoza and Jeffers on Man in Nature', *Inquiry*, 20 (1–4): 481–528.
Slote, M. (1995), 'Agent-Based Virtue Ethics', *Midwest Studies in Philosophy*, 20 (1): 83–101.
Stephens, W. O. (1994). 'Stoic Naturalism, Rationalism, and Ecology', *Environmental Ethics*, 16 (3): 275–86.
Taylor, B. R. (2010), *Dark Green Religion: Nature Spirituality and the Planetary Future*, Berkeley, CA: University of California Press.
Treanor, B. (2014), *Emplotting Virtue: A Narrative Approach to Environmental Virtue Ethics*, Albany, NY: State University of New York Press.
Wenz, P. (2005), 'Synergistic Environmental Virtues: Consumerism and Human Flourishing', in R. D. Sandler and P. Cafaro (eds.), *Environmental Virtue Ethics*, 197–213, Lanham, MD: Rowman & Littlefield.
White, L., Jr. (1967), 'The Historical Roots of Our Ecologic Crisis', *Science*, 155 (3767): 1203–07.

Chapter 5

Banateanu, A. (2001), *La théorie stoïcienne de l'amitié: Essai de reconstruction*, Fribourg: Éditions Universitaires Fribourg Suisse.
Becker, G. K. (2009), 'Je suis le grand tout: Respect for Nature in the Age of Environmental Responsibility', in K.-T. Ip (ed.), *Environmental Ethics: Intercultural Perspectives*, 23–41, Amsterdam and New York: Rodopi.
Bees, R. (2004), *Die Oikeiosislehre der Stoa: I. Rekonstruktion ihres Inhalts*, Würzburg: Königshausen & Neumann.
Bees, R. (2011), 'Kosmische Selbsterhaltung: Soziobiologie und *Oikeiosis*', *Prometheus*, 27: 23–46.
Callicott, J. B. (1989), 'The Conceptual Foundations of the Land Ethic', in *In Defense of the Land Ethic: Essays in Environmental Philosophy*, 75–99, Albany, NY: State University of New York Press.
Celkyte, A. (2017), 'The Stoic Definition of Beauty as *Summetria*', *The Classical Quarterly*, 67: 88–105.
Cheney, J. (1989), 'The Neo-Stoicism of Radical Environmentalism', *Environmental Ethics*, 11: 293–325.
Diehm, C. (2010), 'Minding Nature: Val Plumwood's Critique of Moral Extensionism', *Environmental Ethics*, 32: 3–16.
Gerson, L. P., and B. Inwood (1997), *Hellenistic Philosophy: Introductory Readings*, 2nd edn, Indianapolis, IN and Cambridge: Hackett.
Gill, C. (2006), *The Structured Self in Hellenistic and Roman Thought*, Oxford and New York: Oxford University Press.
Graver, M. R. (2007), *Stoicism and Emotion*. Chicago, IL and London: Chicago University Press.
Henry, P. and H.-R. Schwyzer eds. (1964), Plotinus, *Opera*, vol. 3: *Porphyrii Vita* Plotini; *Enneades I–III*, Oxford: Oxford University Press.
Holland, A. J. (1997), 'Fortitude and Tragedy: The Prospects for a Stoic Environmentalism', in T. M. Robinson and L. Westra (eds.), *The Greeks and the Environment*, 151–66, Lanham, MD et al.: Rowman & Littlefield.

References

Mansfeld, J. (1979), 'Providence and the Destruction of the Universe in Early Stoic Thought', in *Studies in Hellenistic Religions*, 129–88, Leiden: Brill.
Nussbaum, M. C. (2011), *Creating Capabilities: The Human Development Approach*. Cambridge, MA and London: Belknap Press of Harvard University Press.
Panksepp, J., and L. Biven (2012), *The Archaeology of Mind: Neuroevolutionary Origins of Human Emotions*, New York and London: Norton.
Plumwood, V. (1993), *Feminism and the Mastery of Nature*, London and New York: Routledge.
Salles, R., ed. (2009), *God and Cosmos in Stoicism*, New York and Oxford: Oxford University Press.
Sandler, R. L. (2007), *Character and Environment: A Virtue-Oriented Approach to Environmental Ethics*, New York: Columbia University Press.
Seneca (2014), *Hardship & Happiness*, trans. E. Fantham, J. Ker et al., Chicago, IL and London: University of Chicago Press.
Seneca (2015), *Letters on Ethics*, trans. M. R. Graver and A. A. Long, Chicago, IL and London: University of Chicago Press.
Stobaeus (1884), *Ioannis Stobaei Anthologii libri duo priores qui inscribi solent Eclogae physicae et ethicae*, ed. K. Wachsmuth, Berlin: Weidmann.
Stephens, W. O. (1994), 'Stoic Naturalism, Rationalism, and Ecology', *Environmental Ethics*, 16: 275–86.
Taylor, Paul W. (1981), 'The Ethics of Respect for Nature', *Environmental Ethics*, 3: 197–218.
Velayos Castelo, C. (1996), 'Reflections on Stoic Logocentrism', *Environmental Ethics*, 18: 291–96.
Wildberger, J. (2006), *Seneca und die Stoa: Der Platz des Menschen in der Welt*, Berlin and New York: De Gruyter.
Wildberger, J. (2008), 'Beast or God? – The Intermediate Status of Humans and the Physical Basis of the Stoic *scala naturae*', in A. Alexandridis, M. Wild, and L. Winkler-Horaček (eds.), *Mensch und Tier in der Antike: Grenzziehung und Grenzüberschreitung*, 47–70, Wiesbaden: Reichert.
Wildberger, J. (2013), 'Delimiting a Self by God in Epictetus', in J. Rüpke and G. Woolf (eds.), *Religious Dimensions of the Self in the Second Century* CE, 23–45, Tübingen: Mohr Siebeck.
Wildberger, J. (2018), *The Stoics and the State*. Baden Baden: Nomos.
Williams, G. (2012), *A Study of Seneca's 'Natural Questions'*. Oxford: Oxford University Press.

Chapter 6

Attfield, R. (1983), *The Ethics of Environmental Concern*, Oxford: Basil Blackwell.
Attfield, R. (2011), 'Beyond Anthropocentrism', in A. O'Hear (ed.), *Philosophy and the Environment*, Royal Institute of Philosophy Supplement 69, 29–46, Cambridge: Cambridge University Press.
Attfield, R. (2015), *The Ethics of the Global Environment*, 2nd edn., Edinburgh: Edinburgh University Press.
Attfield, R. (2017), *Wonder, Value and God*, London and New York: Routledge.
Belshaw, C. (2001), *Environmental Philosophy: Reason, Nature and Human Concern*, Chesham, UK: Acumen.
Berry, R. J., ed. (2006), *Environmental Stewardship: Critical Perspectives – Past and Present*, London and New York: T&T Clark.
Bookchin, M. (1991), *The Ecology of Freedom*, Montreal: Black Rose Books.
Bratton, S. Power (1988), 'The Original Desert Solitaire: Early Christian Monasticism and Wilderness', *Environmental Ethics*, 10: 31–53.
Brundtland, G. H. ed. (1987), *Our Common Future*, World Commission on Environment and Development, Oxford: Oxford University Press.

References

Callicott, J. B. (1980), 'Animal Liberation: A Triangular Affair', *Environmental Ethics*, 2 (1): 311–38.
Callicott, J. B. (1989), *In Defense of the Land Ethic: Essays in Environmental Philosophy*, Albany, NY: State University of New York Press.
Callicott, J. B. (1999), *Beyond the Land Ethic: More Essays in Environmental Philosophy*, Albany, NY: State University of New York Press.
Carone, G. R. (2001), 'The Classical Greek Tradition', in D. Jamieson (ed.), *A Companion to Environmental Philosophy*, 67–80, Malden, MA and Oxford: Blackwell.
Castelo, C. V. (1996), 'Reflections on Stoic Logocentrism', *Environmental Ethics*, 18: 291–96.
Chadwick, H. (1953), *Origen: Contra Celsum*, Cambridge: Cambridge University Press.
Clark, S. R. L. (1977), *The Moral Status of Animals*, Oxford: Clarendon Press.
Darwin, C. (1985), 'A Historical Sketch', in J. Burrow (ed.), *The Origin of Species*, London: Penguin Classics.
Dodd, C. H. (1938), *The Parables of the Kingdom*, London: Nisbet & Co.
Dubos, R. (1974), 'Franciscan Conservation and Benedictine Stewardship', in D. and E. Spring (eds.), *Ecology and Religion in History*, 114–36, San Francisco and London: Harper & Row.
Gallagher, D. A. and I. J. Gallagher (1966), *Augustine: The Catholic and Manichaen Ways of Life*, Washington: Catholic University of America Press.
Glacken, C. J. (1967), *Traces on the Rhodian Shore: Nature and Culture in Western Thought from Ancient Times to the End of the Eighteenth Century*, Berkeley, CA and London: University of California Press.
Haq, S. N. (2001), 'Islam', in D. Jamieson (ed.), *A Companion to Environmental Philosophy*, 111–29, Malden, MA and Oxford: Blackwell.
Hargrove, E. C. (1989), *Foundations of Environmental Ethics*, Englewood Cliffs, NJ: Prentice-Hall.
Holland, A. (1997), 'Fortitude and Tragedy: The Prospects for a Stoic Environmentalism', in L. Westra and T. M. Robinson (eds.), *The Greeks and the Environment*, 151–66, Lanham, MD and Oxford: Rowman & Littlefield.
Hughes, D. (1994), *Pan's Travail; Environmental Problems of the Ancient Greeks and Romans*, Baltimore and London: Johns Hopkins Press.
Hume, C. W. (1957), *The Status of Animals in the Christian Religion*, London: Universities' Federation for Animal Welfare.
Hursthouse, R. (1999), *On Virtue Ethics*, Oxford: Oxford University Press.
Hursthouse, R. (2002), 'Virtue Ethics vs. Rule-Consequentialism: A Reply to Brad Hooker', *Utilitas*, 14 (1): 41–53.
Johnson, M. R. (2005), *Aristotle on Teleology*, Oxford: Oxford University Press.
Kirk, G. S., J. E. Raven, and M. Schofield, eds. (1983), *The Presocratic Philosophers: A Critical History with a Selection of Texts*, Cambridge: Cambridge University Press.
Leopold, A. (1989), *A Sand County Almanac and Sketches Here and There*, New York: Oxford University Press.
Linzey, A. (1987), *Christianity and the Rights of Animals*, London: SPCK.
Lovejoy, A. O. (1936), *The Great Chain of Being: A Study of the History of an Idea*, Cambridge, MA and London: Harvard University Press.
Lovelock, J. E. (1979), *Gaia: A New Look at Life on Earth*, Oxford and New York: Oxford University Press.
Lovelock, J. E. (2006), *The Revenge of Gaia: Why the Earth is Fighting Back – and How We Can Still Save Humanity*, London and New York: Penguin.
Marlow, H. (2009), *Biblical Prophets and Contemporary Environmental Ethics*, Oxford: Oxford University Press.
Marsh, G. P. [1864] (2003), *Man and Nature*, ed. David Lowenthal, Seattle, WA: University of Washington Press.
Moss, A. W. (1961), *Valiant Crusade*, London: Cassell.

References

Naess, A. (1973), 'The Shallow and the Deep, Long-Range Ecology Movements: A Summary', *Inquiry*, 16: 95–100.
Nagel, E. (1961), *The Structure of Science*, New York: Harcourt, Brace and World.
Nolt, J. (2015), *Environmental Ethics for the Long Term*, London and New York: Routledge.
Nussbaum, M. C. (2006), *Frontiers of Justice: Disability, Nationality, Species Membership*, Cambridge, MA and London: Belknap Press of Harvard University Press.
Passmore, J. (1970), *The Perfectibility of Man*, London: Duckworth.
Passmore, J. (1974), *Man's Responsibility for Nature*, London: Duckworth.
Passmore, J. (1975), 'The Treatment of Animals', *Journal of the History of Ideas*, 36: 145–56.
Preston, C. J. (2009), *Saving Creation: Nature and Faith in the Life of Holmes Rolston III*, San Antonio, TX: Trinity University Press.
Rolston, H., III (1999), *Genes, Genesis and God*, Cambridge: Cambridge University Press.
Santmire, P. (1985), *The Travail of Nature: The Ambiguous Ecological Promise of Christian Theology*, Philadelphia, PA: Fortress Press.
Schofield, M. (1991), *The Stoic Idea of the City*, Cambridge: Cambridge University Press.
Sedley, D. (1991), 'Is Aristotle's Teleology Anthropocentric?', *Phronesis*, 36: 179–96.
Shapiro, H., and E. M. Curley (1965), *Hellenistic Philosophy*, New York: Modern Library.
Shrader-Frechette, K. (2001), 'Ecology', in Dale Jamieson (ed.), *A Companion to Environmental Philosophy*, 304–15, Malden, MA and Oxford: Blackwell.
United Nations, Millennium Development Goals: www.un.org/millenniumgoals/ (accessed 7 May 2018).
United Nations, Sustainable Development Goals: www.un.org/sustainabledevelopment/sustainable-development-goals/ (accessed 8 May 2018).
Waddell, H. [1934] (1995), *Beasts and Saints*, London: Constable.
Wardy, R. (1993), 'Aristotelian Rainfall or the Lore of Averages', *Phronesis*, 38: 18–30.
Webster, Erwin (1923), *Aristotle: Meteorologica*, Oxford: Clarendon Press.
Welchman, J. (2012), 'A Defence of Environmental Stewardship', *Environmental Values*, 21 (3): 297–316.
White, L., Jr. (1967), 'The Historical Roots of Our Ecologic Crisis', *Science*, 155 (37): 1203–07.
Whitney, E. (1993), 'Lynn White, Ecotheology, and History', *Environmental Ethics*, 15: 151–69.
Worrell, R., and M. C. Appleby (2000), 'Stewardship of Natural Resources: Definition, Ethical and Practical Aspects', *Journal of Agricultural and Environmental Ethics*, 12: 263–77.

Chapter 7

Allen, L. C. (1983), *Psalms 101–150*, WBC 21, Waco, TX: Word Books.
Andersen, F. I. (2001), *Habakkuk*, AB 25, New York: Doubleday.
Bauckham, R. (2010), *Bible and Ecology: Rediscovering the Community of Creation*, Sarum Theological Lectures, London: Darton, Longman and Todd.
Beaulieu, M. -C. (2016), *The Sea in the Greek Imagination*, Philadelphia, PA: University of Pennsylvania Press.
Berry, F. (2007), 'Walking on Water: Cosmic Floors in Antiquity and the Middle Ages', *The Art Bulletin*, 89 (4): 627–56.
Berry, R. J., ed. (2006), *Environmental Stewardship: Critical Perspectives – Past and Present*, London: T. & T. Clark.
Boyle, M. O. (2004), '"In the Heart of the Sea": Fathoming the Exodus', *JNES*, 63 (1): 17–27.
Brody, A. J. (1998), *"Each Man Cried out to his God": The Specialized Religion of Canaanite and Phoenician Seafarers*, HSM 58, Atlanta, GA: Scholars Press.

References

Bulbulia, J., et al. (2016), 'To Burn or to Save? The Opposing Functions of Reading Scripture on Environmental Intentions', *Religion, Brain and Behavior*, 6 (4): 278–89.
Clifford, R. J. (1972), *The Cosmic Mountain in Canaan and the Old Testament*, HSM 4, Eugene, OR: Wipf & Stock.
Cole, R. D. (2000), *Numbers*, New American Commentary 3B, Nashville, TN: Broadman and Holman.
Day, J. (1985), *God's Conflict with the Dragon and the Sea: Echoes of a Canaanite Myth in the Old Testament*, Cambridge: Cambridge University Press.
Dijk, H. J. van (1968), *Ezekiel's Prophecy on Tyre (Ez. 26,1 – 28, 19): A New Approach*, BibOr 20, Rome: Pontifical Biblical Institute.
Dillmann, A. (1897), *Genesis Critically and Exegetically Expounded, Volume 1*, trans. from 6th German edition (1892) by W. B. Stevenson, Edinburgh: T. & T. Clark.
Geyer, J. B. (2004), *Mythology and Lament: Studies in the Oracles about the Nations*, SOTSMS, Aldershot: Ashgate.
Haak, R. D. (1992), *Habakkuk*, VTSup 44, Leiden: E.J. Brill.
Horowitz, W. (1998), *Mesopotamian Cosmic Geography*, Winona Lake, IN: Eisenbrauns.
Horrell, D. G. (2010), *The Bible and the Environment: Towards a Critical Ecological Biblical Theology*, London: Equinox.
Hummel, H. D. (2007), *Ezekiel 21–48*, Concordia Commentary, Saint Louis, MO: Concordia.
Humphreys, C. J. (1998), 'The Number of People in the Exodus from Egypt: Decoding Mathematically the Very Large Numbers in Numbers I and XXVI', *VT*, 48 (2): 196–213.
Humphreys, C. J. (2000), 'The Numbers in the Exodus from Egypt: A Further Reappraisal', *VT*, 50 (3): 323–28.
Katzenstein, H. J. (1973), *The History of Tyre*, Jerusalem: Shocken.
Keel, O. (1997), *The Symbolism of the Biblical Word: Ancient Near Eastern Iconography and the Book of Psalms*, Winona Lake, IN: Eisenbrauns.
Kraus, H. -J. (1993), *Psalms 60-150: A Continental Commentary*, trans. from 6th German edition (1978) by H. C. Oswald, Minneapolis: Fortress Press.
Lee, L. (2016), *Mapping Judah's Fate in Ezekiel's Oracles against the Nations*, Ancient Near East Monographs 15, Atlanta, GA: SBL Press.
Levenson, J. D. (1976), *Theology of the Program of Restoration of Ezekiel 40–48*, Missoula, MT: Scholars Press for Harvard Semitic Museum.
Levine, B. A. (1993), *Numbers 1–20: A New Translation with Introduction and Commentary*, AB 4A, New York: Doubleday.
Louth, A., ed., in collaboration with M. Conti (2001), *Genesis 1-11*, Ancient Christian Commentary on Scripture: Old Testament, I; Downers Grove, IL: InterVarsity Press.
Marlow, H. (2009), *Biblical Prophets and Contemporary Environmental Ethics: Re-Reading Amos, Hosea, and First Isaiah*, Oxford: Oxford University Press.
Morales, L. M., ed. (2014), *Cult and Cosmos: Tilting Toward a Temple-Centered Theology*, Biblical Tools and Studies 18, Leuven: Peeters.
Oppenheim, A. L. (1977), *Ancient Mesopotamia*, rev. edn completed by E. Reiner, Chicago, IL: University of Chicago Press.
Pope, M. H. (1955), *El in the Ugaritic Texts*, VTSup 2, Leiden: E. J. Brill.
Richards, I. A. (1936), *The Philosophy of Rhetoric*, London: Oxford University Press.
Rodd, C. S. (2001), *Glimpses of a Strange Land: Studies in Old Testament Ethics*, Edinburgh: T. & T. Clark.
Seybold, K. (1996), *Die Psalmen*, HAT 1 (15), Tübingen: J.C.B. Mohr.
Von Rad, G. (1972), *Genesis: a Commentary*, trans. from 9th German edition by J. H. Marks with revisions by J. Bowden, London: SCM Press.

References

Watson, R. S. (2005), *Chaos Uncreated: A Reassessment of the Theme of "Chaos" in the Hebrew Bible*, BZAW 341, Berlin: De Gruyter.
Wenham, G. T. (1985), 'Sanctuary Symbolism in the Garden of Eden Story' in *Proceedings of the World Congress of Jewish Studies*, 9: 19–25.
Westermann, C. (1988), *Genesis*, trans. D. E. Orton, Edinburgh: T. & T. Clark.
White, L., Jr. (1967), 'The Historical Roots of our Ecologic Crisis', *Science*, 155 (3767): 1203–07.
Wilson, I. D. (2013), 'Tyre, A Ship: The Metaphorical World of Ezekiel 27 in Ancient Judah', *ZAW*, 125 (2): 249–62.
Wright, C. J. H. (2004), *Old Testament Ethics for the People of God*, Leicester: InterVarsity Press.
Wyatt, N. (2001), *Space and Time in the Religious Life of the Near East*, Sheffield: Sheffield Academic Press.
Yoder, T. R. (2016), *Fishers of Fish and Fishers of Men: Fishing Imagery in the Hebrew Bible and the Ancient Near East*, Explorations in Ancient Near Eastern Civilizations 4, Winona Lake, IN: Eisenbrauns.
Zimmerli, W. (1983), *Ezekiel*, II, trans. J. D. Martin, Philadelphia, PA: Fortress.

Chapter 8

Anhalt, E. K. (1993), *Solon the Singer: Politics and Poetics*, Lanham, MD: Rowman and Littlefield.
Bacon, H. (2001), 'The Furies' Homecoming', *CPh*, 96: 48–59.
Bakola, E. (2014), 'Interiority, the "Deep Earth", and the Spatial Symbolism of Darius' Apparition in the Persians of Aeschylus', *CCJ*, 60: 1–36.
Bakola, E. (2016), 'Textile Symbolism and the "Wealth of the Earth": Creation, Production and Destruction in the "Tapestry Scene" of Aeschylus' *Oresteia* (*Ag.* 905-78)', in M. Harlow, M.-L. Nosch, and G. Fanfani (eds.), *Spinning Fates and the Song of the Loom: the Use of Textiles, Clothing and Cloth Production as Metaphor, Symbol and Narrative*, 115–36, Oxford: Oxbow Books.
Bakola, E. (2018), 'Seeing the Invisible: Interior Spaces and Uncanny Erinyes in Aeschylus' *Oresteia*', in A. Kampakoglou and A. Novokhatko (eds.), *Gaze, Vision and Visuality in Ancient Greek Literature*, 163–86, Berlin: De Gruyter.
Balot, R. K. (2001), *Greed and Injustice in Classical Athens*, Princeton, NJ and Oxford: Princeton University Press.
Boedeker, D. D. (1984), *Descent from Heaven: Images of Dew in Greek Poetry and Religion*, Chico, CA: Scholars Press.
Boys-Stones, G. R., and J. H. Haubold, eds. (2010), *Plato and Hesiod*, Oxford: Oxford University Press.
Bremmer, J. N. (1994), *Greek Religion (New Surveys in the Classics No. 24)*, Cambridge: Cambridge University Press.
Brown A. L. (1984), 'Eumenides in Greek Tragedy', *Classical Quarterly*, 34: 260–81.
Burkert, W. (1985), *Greek Religion: Aarchaic and Classical*, trans. J. Raffan, Oxford: Basil Blackwell.
Catenaccio, C. (2011), 'Dream as Image and Action in Aeschylus' *Oresteia*', *Greek, Roman, and Byzantine Studies* 51: 202–31.
Chantraine, P. (1968), *Dictionnaire étymologique de la langue grecque: histoire des mots*, Paris: Klincksieck.
Clay, J. S. (2003), *Hesiod's Cosmos*, Cambridge: Cambridge University Press.
Collard, C. (2003), *Aeschylus Oresteia, translated with an introduction and notes*, Oxford: Oxford University Press.
Daraki, M. (1999), *Dionysos et la déesse Terre*, Paris: Flammarion.

References

Dawe, R. D. (1964), *The Collation and Investigation of Manuscripts of Aeschylus*, Cambridge: Cambridge University Press.

Deacy, S. (2015), 'Gods – Olympian or Chthonian?' in E. Eidinow and J. Kindt (eds.), *The Oxford Handbook of Ancient Greek Religion*, 355–67, Oxford: Oxford University Press.

Dowden, K. (2007), 'Olympian Gods, Olympian Pantheon', in D. Ogden (ed.), *A Companion to Greek Religion*, 41–55, Oxford: Blackwell.

Easterling, P. (2008), 'Theatrical Furies: Thoughts on Eumenides', in M. Revermann and P. Wilson (eds.), *Performance, Iconography, Reception: Studies in Honour of Oliver Taplin*, 219–36, Oxford: Oxford University Press.

Ekroth, G. (2002), *The Sacrificial Rituals of Greek Hero-Cults in the Archaic to the Hellenistic Periods*, Liège: Presses universitaires de Liège.

Felton, D. (2007), 'The Dead', in D. Ogden (ed.), *A Companion to Greek Religion*, 86–99, Oxford: Blackwell.

Fischer, A. (2010), 'Chthonic Gods', in M. Gagarin and E. Fantham (eds.), *The Oxford Encyclopedia to Ancient Greece and Rome*, 115–16, Oxford: Oxford University Press.

Fowler, B. (1991), 'The Creatures and the Blood', *ICS*, 16: 85–100.

Fraenkel, E. (1950), *Aeschylus: Agamemnon*, vol. II, Oxford: Oxford University Press.

Furley, W. D. (1986), 'Motivation in the Parodos of Aeschylus' *Agamemnon*', *CPh*, 81: 109–21.

Garvie, A. F. (1986), *Aeschylus Choephori*, with introduction and commentary, Oxford: Clarendon Press.

Goheen, R. F. (1955), 'Aspects of Dramatic Symbolism: Three Studies in the *Oresteia*', *American Journal of Philology*, 76: 113–37.

Goldhill, S. (1986), *Reading Greek Tragedy*, Cambridge: Cambridge University Press.

Goldhill, S. (2004), *Aeschylus: The Oresteia* (first edn. 1992), Cambridge: Cambridge University Press.

Gregory, A. (2016), *Anaximander: A Reassessment*, London: Bloomsbury.

Hadzisteliou-Price, T. H. (1978), *Kourotrophos: Cults and Representations of the Greek Nursing Deities*, Leiden: Brill.

Hägg, R., and B. Alroth, eds. (2005), *Greek Sacrificial Ritual, Olympian and Chthonian. Proceedings of the Sixth International Seminar on Ancient Greek Cult, organized by the Department of Classical Archaeology and Ancient History, Göteborg University, 25–27 April 1997*, Stockholm: Åströms Förlag.

Heath, J. (1999), 'Disentangling the Beast: Humans and Other Animals in Aeschylus' *Oresteia*', *Journal of Hellenic Studies*, 119: 17–47.

Helm J. J. (1993), '*Koros*: from Satisfaction to Greed', *Classical World*, 87: 5–11.

Henrichs, A. (1991), 'Namenlosigkeit und Euphemismus: zur Ambivalenz der chthonischen Mächte im attischen Drama', in H. Hofmann and A. Harder (eds.), *Fragmenta Dramatica: Beiträge zur Interpretation der griechischen Tragikerfragmente und ihrer Wirkungsgeschichte*, 161–201, Göttingen: Vandenhoeck & Ruprecht.

Hunter, R. (2014), *Hesiodic Voices. Studies in the Ancient Reception of Hesiod's Works and Days*, Cambridge: Cambridge University Press.

Hutchinson, G. O. (1985), *Aeschylus: Septem contra Thebas*, edited with introduction and commentary, Oxford: Clarendon Press.

Iles Johnston, S. (1999), *Restless Dead: Encounters Between the Lliving and the Dead in Ancient Greece*, Berkeley, CA and London: University of California Press.

Jones, J. (1962), *On Aristotle and Greek Tragedy*, London: Chatto & Windus.

Kearns, E. (1989), *The Heroes of Attica*, Bulletin of the Institute of Classical Studies Supp. 57, London: Institute of Classical Studies.

Labarrière, J. -L. et al., eds. (2006), *Avez-vous vu les Erinyes?*, Mètis 4, Paris and Athens: Éditions de l'École des hautes études en sciences sociales.

References

Lardinois, A. (1992), 'Greek Myths for Athenian Rituals: Religion and Politics in Aeschylus' *Eumenides* and Sophocles' *Oedipus Coloneus*', *Greek, Roman, and Byzantine Studies*, 33: 313–25.
Lebeck, A. (1971), *The Oresteia*, Washington, DC: Center for Hellenic Studies.
Lloyd-Jones, H. (1971), *The Justice of Zeus*, Berkeley, CA and London: University of California Press.
Lloyd-Jones, H. (1990), 'Erinyes, Semnai Theai, Eumenides', in E. Craik (ed.), *Owls to Athens: Essays on Classical Subject Presented to Sir Kenneth Dover*, 203–11, Oxford: Clarendon Press.
Loraux, N. (1993), *The Children of Athena: Athenian Ideas about Citizenship and the Division between the Sexes*, trans. C. Levine, Princeton, NJ: Princeton University Press.
Mitchell-Boyask, R. (2009), *Aeschylus: Eumenides*, London: Bloomsbury.
Nagy G. (1985), 'Theognis and Megara: A Poet's Vision of His City', in T. J. Figueira and G. Nagy (eds.), *Theognis of Megara: Poetry and the Polis*, 22–81, Baltimore, MD and London: Johns Hopkins University Press.
Nelson, S. A. (1998), *God and the Land: The Metaphysics of Farming in Hesiod and Vergil*, New York and Oxford: Oxford University Press.
Noussia-Fantuzzi, M. (2015), 'Solon in Tragedy', in G. Nagy and M. Noussia-Fantuzzi (eds.), *Solon in the Making: The Early Reception in the Fifth and Fourth Centuries*, 43–65, Berlin and Boston, MA: De Gruyter.
Padel, R. (1992), *In and Out of the Mind: Greek Images of the Tragic Self*, Princeton, NJ: Princeton University Press.
Page, D. (1972), *Aeschylus, Septem quae supersunt tragoedias*, Oxford: Clarendon Press.
Parke, H. W. (1977), *Festivals of the Athenians*, London: Thames & Hudson.
Parker R. (2005), *Polytheism and Society at Athens*, Oxford: Oxford University Press.
Parker, R. (2009), 'Aeschylus' Gods: Drama, Cult, Theology', in M. Griffith, et al. (eds.), *Eschyle à l'aube du théâtre occidental: neuf exposés suivis de discussions: Vandœuvres-Genève, 25-29 août 2008*, 127–54, Geneva: Fondation Hardt.
Parker, R. (2011), *On Greek Religion*, Ithaca, NY and London: Cornell University Press.
Peradotto, J. J. (1964), 'Some Patterns of Nature Imagery in the *Oresteia*', *American Journal of Philology*, 85: 378–93.
Raeburn, D., and O. Thomas (2011), *The Agamemnon of Aeschylus*, Oxford: Oxford University Press.
Rehm, R. (2002), *The Play of Space: Spatial Transformation in Greek Tragedy*, Princeton, NJ and Oxford: Princeton University Press.
Roisman, H. M., ed. (2014), *The Encyclopedia of Greek Tragedy*, Chichester: Wiley-Blackwell.
Rosenbloom, D. (2006), *Aeschylus: Persians*, London: Duckworth.
Rosenmeyer, T. G. (1982), *The Art of Aeschylus*, Berkeley, CA: University of California Press.
Schlesier, R. (1991/92), 'Olympian versus Chthonian Religion', *Scripta Classica Israelica*, 11: 38–51.
Schlesier, R. (1997), 'Chthonische Götter', in H. Cancik, H. Schneider, and M. Landfester (eds.), *Der Neue Pauly* II, 1185–90, Stuttgart: J.B. Mettzler.
Scodel, R. (1996), '*Domon agalma*: Virgin Sacrifice and Aesthetic Object', *Transactions of the American Philological Association*, 126: 111–28.
Scott, W. C. (1966), 'Wind Imagery in the *Oresteia*', *Transactions of the American Philological Association*, 97: 459–71.
Scullion, S. (1994), 'Olympian and Chthonian', *Classical Antiquity*, 13: 75–119.
Scullion, S. (2005), ' "Saviours of the Father's Hearth": Olympian and Chthonian in the *Oresteia*', in R. Hägg and B. Alroth (eds.), *Greek Sacrificial Ritual, Olympian and Chthonian. Proceedings of the Sixth International Seminar on Ancient Greek Cult, organized by the Department of Classical Archaeology and Ancient History, Göteborg University, 25–27 April 1997*, 23–36, Stockholm: Åströms Förlag.
Seaford, R. (2013), *Cosmology and the Polis: The Social Construction of Space and Time in the Tragedies of Aeschylus*, Cambridge: Cambridge University Press.

References

Segal, C. (1998), *Sophocles' Tragic World: Divinity, Nature, Society*, Cambridge, MA and London: Harvard University Press.
Sewell-Rutter, N. J. (2007), *Guilt by Descent: Moral Inheritance and Decision Making in Greek Tragedy*, Oxford: Oxford University Press.
Solmsen, F. (1949), *Hesiod and Aeschylus*, Ithaca, NY: Cornell University Press.
Sommerstein, A. H. (1989), *Aeschylus, Eumenides*, Cambridge: Cambridge University Press.
Sommerstein, A. H. (2008), *Aeschylus*, Cambridge, MA and London: Harvard University Press.
Sommerstein, A. H. (2010), *Aeschylean Tragedy* (first edn. 1996, Bari), London: Duckworth.
Stamatopoulou, Z. (2017), *Hesiod and Classical Greek Poetry: Reception and Transformation in the Fifth Century BCE*, Cambridge: Cambridge University Press.
Stokes, M. C. (1962), 'Hesiodic and Milesian Cosmogonies I', *Phronesis*, 7: 1–37.
Stokes, M. C. (1963), 'Hesiodic and Milesian Cosmogonies: II', *Phronesis*, 8: 1–34.
Taplin, O. (1977), *The Stagecraft of Aeschylus: The Dramatic Use of Exits and Entrances in Greek Tragedy*, Oxford: Clarendon Press.
Van Noorden, H. (2015), *Playing Hesiod: The 'Myth of the Races' in Classical Antiquity*, Cambridge: Cambridge University Press.
Wilamowitz-Moellendorff, U. von (1931), *Der Glaube der Hellenen*, vol. 1, Berlin: Weidmannsche buchhandlung.
Winnington-Ingram, R. P. (1983), *Studies in Aeschylus*, Cambridge: Cambridge University Press.
Zeitlin, F. I. (1978), 'The Dynamics of Misogyny: Myth and Mythmaking in the *Oresteia* of Aeschylus', *Arethusa*, 11: 149–84; repr. in J. Peradotto and J. P. Sullivan, eds. (1984), *Women in the Ancient World: The Arethusa Papers*, 159–94, Albany, NY: State University of New York Press; also repr. in F. I. Zeitlin (1996), *Playing the Other: Gender and Society in Classical Greek Literature*, 87–122, Chicago, IL: University of Chicago Press.
Zografou, A. (2010), *Chemins d'Hécate. Portes, routes, carrefours et autres figures de l'entre-deux*, Liège: Presses universitaires de Liège.

Chapter 9

Bausinger, T., E. Bonnaire, and J. Preuss (2007), 'Exposure Assessment of a Burning Ground for Chemical Ammunition on the Great War Battlefields of Verdun', *Science of the Total Environment*, 382 (2–3): 259–71.
Biddle, M. E. (1996), *Polyphony and Symphony in Prophetic Literature: Rereading Jeremiah 7-20*, Macon, GA: Mercer University Press.
Blake, William (1993 [1810]), *Milton a Poem: And the Final Illuminated Works*, eds. R. N. Essick et al., London: William Blake Trust/Tate Gallery).
Borden, M. (2008), *The Forbidden Zone*, ed. H. Hutchison, London: Hesperus.
Borden, M. (2015), *Poems of Love and War*, ed. P. O'Prey, London: Dare-Gale Press.
Bottomley, G. and D. Harding, eds. (1977), *The Collected Poems of Isaac Rosenberg*, London: Chatto & Windus.
Carroll, R, M. D. (1992), *Contexts for Amos: Prophetic Poetics in Latin American Perspective*, Sheffield: Sheffield Academic Press.
Childs, B. S. (1959), 'The Enemy From the North and the Chaos Tradition', *JBL*, 78: 187–98.
Copp, M., ed. (2001), *Cambridge Poets of the Great War: An Anthology*, Madison, N.J.; London: Fairleigh Dickinson University Press; Associated University Press.
Craigie, P., P. H. Kelley, and J. F. Drinkard Jr. (1991), *Jeremiah 1-25*, WBC 26, Dallas, TX: Word Books.

References

de Miroschedji, P. (2012), 'Egypt and Southern Canaan in the Third Millennium BCE: Uni's Asiatic Campaign Revisited', in M. I. Gruber, et al. (eds.), *All the Wisdom of the East: Studies in Near Eastern Archaeology and History in Honor of Eliezer D. Oren*, 265–93, Fribourg; Göttingen: Academic Press; Vandenhoeck & Ruprecht.

DeRoche, M. (1980), 'Zephaniah I 2-3: The "Sweeping" of Creation', *VT*, 30 (3): 104–8.

DeRoche, M. (1981), 'The Reversal of Creation in Hosea', *VT*, 31: 401–9.

Donin, H. H. (1980), *To Pray as a Jew: A Guide to the Prayer Book and the Synagogue Service*, New York: Basic Books.

Dubois, J. -J. (1994), 'Les Forêts de Frontière du Nord de la France', in A. Corvol and J. -P. Amat (eds.), *Forêt et guerre*, 165–81, Paris: L'Harmattan.

Eberhart, R., and S. Rodman, eds. (1945), *War and the Poet: An Anthology of Poetry Expressing Man's Attitudes to War from Ancient Times to the Present*, New York: Devin-Adair.

Ferris, P. W. (1992), *The Genre of Communal Lament in the Bible and the Ancient Near East*, Atlanta, GA: Scholars Press.

Fretheim, T. E. (2005), *God and World in the Old Testament: A Relational Theology of Creation*, Nashville, TN: Abingdon.

Fussell, P. (2000), *The Great War and Modern Memory*, Oxford: Oxford University Press.

Golden, J. M. (2004), *Ancient Canaan and Israel: New Perspectives*, Santa Barbara, CA and; Oxford: ABC-CLIO.

Goodhand, J. (2003), 'Enduring Disorder and Persistent Poverty: A Review of the Linkages Between War and Chronic Poverty', *World Development*, 31 (3): 629–46.

Habel, N., ed. (2000), *Readings from the Perspective of the Earth*, Sheffield: Sheffield Academic Press.

Habel, N., and P. Trudinger, eds. (2008), *Exploring Ecological Hermeneutics*, Atlanta, GA: Society of Biblical Literature.

Hayes, K. M (2002), *'The Earth Mourns': Prophetic Metaphor and Oral Aesthetic*, SBLABib, 8; Atlanta, GA: Society of Biblical Literature.

Henderson, J. M. (2007), 'Jeremiah 2-10 as a Unified Literary Composition: Evidence of Dramatic Portrayal and Narrative Progression', in J. Goldingay (ed.), *Uprooting and Planting: Essays on Jeremiah for Leslie Allen*, LHBOTS 459, 116–52, New York and London: T & T Clark.

Herzog, C., and M. Gichon (1978), *Battles of the Bible*, London: Weidenfeld and Nicolson.

Holladay, W. L. (1961), 'On Every High Hill and Under Every Green Tree', *VT*, 11 (2): 170–6.

Housman, A. E. (1988), *Collected Poems and Selected Prose*, London: Penguin.

Hughes, J. D. (2013), 'Warfare and Environment in the Ancient World', in B. Campbell and A. T. Lawrence (eds.), *The Oxford Handbook of Warfare in the Classical World*, Oxford: Oxford University Press.

Hughes, J. D., and J. V. Thirgood (1982), 'Deforestation, Erosion, and Forest Management in Ancient Greece and Rome', *Journal of Forest History*, 26 (2): 60–75.

Hupy, J. P., and R. J. Schaetzl (2008), 'Soil Development on the WWI Battlefield of Verdun, France', *Geoderma*, 145 (1–2): 37–49.

Kendall, T. (2014), *Poetry of the First World War: An Anthology*, Oxford: Oxford University Press.

Laub, D. (1995), 'Truth and Testimony: The Process and the Struggle', in C. Caruth (ed.), *Trauma: Explorations in Memory*, Baltimore, MD and London: Johns Hopkins University Press, 61–75.

Lee, S. (1996), 'Introduction to First World War Poetry', The First World War Poetry Digital Archive. http://projects.oucs.ox.ac.uk/jtap/tutorials/intro/intro.html (accessed 20 June 2018).

Leitenberg, M. (2006), *Deaths in Wars and Conflicts in the 20th Century*, Ithaca, NY: Cornell University.

Marlow, H. (2009), *Biblical Prophets and Contemporary Environmental Ethics: Re-reading Amos, Hosea and First Isaiah*, Oxford: Oxford University Press.

Marlow, H. (2012), 'Land', in M. J. Boda and J. G. McConville (eds.), *Dictionary of the Old Testament Prophets*, 489–93, Downers Grove, IL and Nottingham: InterVarsity Press.

Marlow, H. (2013), 'The Hills are Alive! The Personification of Nature in the Psalter', in D. A. Baer and R. P. Gordon (eds.), *Leshon Limmudim: Essays on the Language and Literature of the Hebrew Bible in Honour of A.A. Macintosh*, 189–203, London: Bloomsbury.
Martin, P. S., and C. R. Szuter (1999), 'War Zones and Game Sinks in Lewis and Clark's West', *Conservation Biology*, 13 (1): 36–45.
Mowinckel, S. (1962), *The Psalms in Israel's Worship I*, trans. D. R. Ap-Thomas, Oxford: Basil Blackwell.
Nichols, R. (1917), *Ardours and Endurances*, New York: Frederick H. Stokes.
O'Connor, K. M. (1988), *The Confessions of Jeremiah: Their Interpretation and Role in Chapters 1-25*, Atlanta, GA: Scholars Press.
O'Connor, K. M. (2011), *Jeremiah: Pain and Promise*, Minneapolis, MN: Fortress Press.
Puyo, J. -Y. (2004), 'Les conséquences de la Première Guerre mondiale pour les forêts et les forestiers français', *Revue Forestière Française*, 56 (6): 573–84.
Reilly, C., ed. (1981), *Scars Upon My Heart: Women's Poetry and Verse of the First World War*, London: Virago.
Sassoon, S. (1984), *Collected Poems 1908-1956*, London and Boston, MA: Faber and Faber.
Silkin, J., ed. (1979), *The Penguin Book of First World War Poetry*, London: Allen Lane.
Simkins, R. (1991), *Yahweh's Activity in History and Nature in the Book of Joel*, Lewiston, NY: Edwin Mellen Press.
Smith-Christopher, D. (2011), 'Reading War and Trauma: Suggestions toward a Social-Psychological Exegesis of Exile and War in Biblical Texts', in J. L. Wright, B. E. Kelle, and F. R. Ames (eds.), *Interpreting Exile: Displacement and Deportation in Biblical and Modern Contexts*, 253–76, Atlanta, GA: Society of Biblical Literature.
Stallworthy, J. (2014), *The New Oxford Book of War Poetry*, Oxford: Oxford University Press.
Stulman, L. (2005), *Jeremiah*, Abingdon Old Testament Commentaries, Nashville, TN: Abingdon Press.
Tate, T. (2009), 'The First World War: British Writing', in C. M. McLoughlin (ed.), *The Cambridge Companion to War Writing*, 160–74, Cambridge and New York: Cambridge University Press.
Theisen, O. M., N. P. Gleditsch, and H. Buhaug (2013), 'Is Climate Change a Driver of Armed Conflict', *Climatic Change*, 117 (3): 613–25.
Van Meirvenne, M., et al. (2008), 'Could Shelling in the First World War Have Increased Copper Concentrations in the Soil around Ypres?', *European Journal of Soil Science*, 59 (2): 372–79.
Vandiver, E. (2010), *Stand in the Trench, Achilles: Classical Receptions in British Poetry of the Great War*, Oxford: Oxford University Press.

Chapter 10

Attfield, R. (1983), *The Ethics of Environmental Concern*, Oxford: Blackwell.
Attfield, R. (1994), *Environmental Philosophy: Principles and Prospects*, Aldershot: Avebury.
Bakola, E. (forthcoming) *The Erinyes and the Wealth of the Earth: Cosmos, Nature and Resources in Aeschylean Tragedy*.
Balabanski, V. (2010), 'Hellenistic Cosmology and the Letter to the Colossians: Towards an Ecological Hermeneutic', in D. G. Horrell et al. (eds.), *Ecological Hermeneutics: Biblical, Historical and Theological Perspectives*, 94–107, London: T & T Clark.
Barr, J. (1972), 'Man and Nature – the Ecological Controversy and the Old Testament', *Bulletin of the John Rylands Library*, 55 (1): 9–32.
Bauckham, R. (2010), *The Bible and Ecology: Rediscovering the Community of Creation*, Waco, TX: Baylor University Press.
Berry, T. (1998), *The Dream of the Earth*, San Francisco, CA: Sierra Club Books.

References

Birge, D. (1994), 'Trees in the Landscape of Pausanias' Periegesis', in S. E. Alcock and R. Osborne (eds.), *Placing the Gods: Sanctuaries and Sacred Space in Ancient Greece*, 231–46, Oxford: Clarendon.

Borlik, T. A. (2011), *Ecocriticism and Early Modern English Literature: Green Pastures*, London: Routledge.

Cole, S. G. (2004), *Landscapes, Gender, and Ritual Space: The Ancient Greek Experience*, Berkeley, CA: University of California Press.

Collins, P. (1995), *God's Earth: Religion as if Matter Really Mattered*, Dublin: Gill & Macmillan.

Conradie, E. (2010), 'Towards a Theological Ecological Hermeneutic', in D. G. Horrell et al. (eds.), *Ecological Hermeneutics: Biblical, Historical and Theological Perspectives*, London: T & T Clark.

Cox, H. (1967), *The Secular City: Secularization and Urbanization in Theological Perspective*, London: SCM Press.

Dillon, M. P. J. (1997), 'The Ecology of the Greek Sanctuary', *ZPE*, 118: 113–27.

Dowden, K. (2000), *European Paganism: The Realities of Cult from Antiquity to the Middle Ages*, London: Routledge.

Fox, M. (1983), *Original Blessing*, Santa Fe, NM: Bear.

Gellner, D. N. (1999), 'Anthropological Approaches', in P. Connolly (ed.), *Approaches to the Study of Religion*, 10–41, London: Continuum.

Glacken, C. J. (1967), *Traces on the Rhodian Shore: Nature and Culture in Western Thought from Ancient Times to the End of the Eighteenth Century*, Berkeley, CA: University of California Press.

Habel, N. C. (2000), *Readings from the Perspective of Earth*, Sheffield: Sheffield Academic Press.

Habinek, T. N. (1998), *The Politics of Latin Literature: Writing, Identity and Empire in Ancient Rome*, Princeton, NJ: Princeton University Press.

Harris, A. (2013), 'Embodied Eco-Paganism', in G. Harvey (ed.), *The Handbook of Contemporary Animism*, 403–15, Durham: Acumen.

Harris, M. (1968), *The Rise of Anthropological Theory: A History of Theories of Culture*, London: Routledge and Kegan Paul.

Harvey, G. (1997), *Contemporary Paganism: Listening People, Speaking Earth*, Washington Square, NY: New York University Press.

Harvey, G. (2009), 'Animist Paganism', in J. R. Lewis and M. Pizza (eds.), *Handbook of Contemporary Paganism*, 393–411, Leiden: Brill.

Harvey, G. (2013), 'Introduction', in G. Harvey (ed.), *The Handbook of Contemporary Animism*, 1–12, Durham: Acumen.

Henrichs, A. (1979), '"Thou Shalt not Kill a Tree": Greek, Manichean and Indian Tales', *BASP*, 16: 35–108.

Hiebert, T. (1996), *The Yahwist's Landscape: Nature and Religion in Early Israel*, Oxford: Oxford University Press.

Hodgson, P. C., ed. (1987), *Hegel: Lectures on the Philosophy of Religion*, vol. 2, Berkeley, CA: University of California Press.

Hogan, L. (2013), 'We call it tradition', in G. Harvey (ed.), *The Handbook of Contemporary Animism*, 17–26, Durham: Acumen.

Horrell, D. (2010), *The Bible and the Environment*, London: Equinox.

Horrell, D., C. Hunt, and C. Southgate, eds. (2010), *Greening Paul: Rereading the Apostle in a Time of Ecological Crisis*, Waco, TX: Baylor University Press.

Horrell, D. G. et al. (2010), *Ecological Hermeneutics: Biblical, Historical and Theological Perspectives*, London: T & T Clark.

Hughes, J. D. (2014), *Environmental Problems of the Greeks and Romans: Ecology in the Ancient Mediterranean*, Baltimore, MD: Johns Hopkins University Press.

References

Hunt, A. (2016), *Reviving Roman Religion: Sacred Trees in the Roman World*, Cambridge: Cambridge University Press.

Jantzen, G. (1984), *God's World, God's Body*, London: Darton, Longman and Todd.

Jedan, C. (2017), 'A Different Kind of Reformation: Revisiting the Lynn White Thesis', *NTT: Journal for Theology and the Study of Religion*, 71 (3): 277–86.

Jenkins, W. (2009), 'After Lynn White: Religious Ethics and Environmental Problems', *The Journal of Religious Ethics*, 37 (2): 283–309.

Lamond, F. (1997), *Religion Without Beliefs: Essays in Pantheist Theology, Comparative Religion and Ethics*, London: Janus.

Lane Fox, R. (1986), *Pagans and Christians*, Harmondsworth: Viking.

Larson, J. (2010), 'A Land Full of Gods: Nature Deities in Greek Religion', in D. Ogden (ed.), *A Companion to Greek Religion*, 56–70, Oxford: Wiley Blackwell.

Lessa, W. A., and E. Z. Vogt (1972), 'General Introduction', in W. A. Lessa and E. Z. Vogt (eds.), *Reader in Comparative Religion: An Anthropological Approach*, 1–6, New York: Harper and Row.

LeVasseur, T., and A. Peterson (2017), *Religion and Ecological Crisis: The "Lynn White Thesis" at Fifty*, New York: Routledge.

Lovelock, J. E. (1990), 'A Danger to Science?', *Nature*, 348 (6303): 685.

McFague, S. (1993), *The Body of God: An Ecological Theology*, London: SCM.

Marlow, H. (2009), *Biblical Prophets and Contemporary Environmental Ethics: Re-Reading Amos, Hosea and First Isaiah*, Oxford: Oxford University Press.

Mikalson, J. D. (2010), *Ancient Greek Religion*, 2nd edn., Chichester: Wiley-Blackwell.

Moltmann, J. (1985), *God in Creation: An Ecological Doctrine of Creation*, London: SCM.

Northcott, M. S. (1996), *The Environment and Christian Ethics*, Cambridge: Cambridge University Press.

Passmore, J. A. (1974), *Man's Responsibility for Nature: Ecological Problems and Western Traditions*, London: Duckworth.

Rigby, K. (2002), 'Ecocriticism', in J. Wolfreys (ed.), *Introducing Criticism at the 21st century*, 151–78, Edinburgh: Edinburgh University Press.

Rives, J. B. (2007), *Religion in the Roman Empire*, Oxford: Blackwell.

Rose, H. J. (1948), *Ancient Roman Religion*, London: Hutchinson's University Library.

Rudd, G. (2014), 'Being Green in Late Medieval English Literature', in G. Garrard (ed.), *The Oxford Handbook of Ecocriticism*, 78–97, Oxford: Oxford University Press.

Sallares, R. (1991), *The Ecology of the Ancient Greek World*, London: Duckworth.

Santmire, H. P. (1985), *The Travail of Nature: The Ambiguous Ecological Promise of Christian Theology*, Minneapolis, MN: Fortress Press.

Scheid, J. (1987), 'Polytheism Impossible; Or, the Empty Gods: Reasons Behind a Void in the History of Roman Religion', *History and Anthropology*, 3: 303–25.

Schwarz, W. (1991), 'The Rebirth of Mother Earth', *The Guardian*, 07 January 1991: 33.

Sharpe, E. J. (1975), *Comparative Religion: A History*, London: Duckworth.

Sheldrake, R. (1990), *The Rebirth of Nature: The Greening of Science and God*, London: Century.

Sideris, L. (2006), 'Religion, Environmentalism and the Meaning of Ecology', in R. Gottlieb (ed.), *The Oxford Handbook of Religion and Ecology*, 446–64, Oxford: Oxford University Press.

Sourvinou-Inwood, C. (1990), 'What is Polis Religion?', in O. Murray and S. R. F. Price (eds.), *The Greek City: From Homer to Alexander*, 295–322, Oxford: Clarendon Press.

Stocking, G. W. (1987), *Victorian Anthropology*, Oxford: Maxwell Macmillan.

Stone, C. D. (1972), 'Should Trees Have Standing? Toward Legal Rights for Natural Objects', *Southern California Law Review*, 45: 450–87.

Thomas, R. F. (1988), 'Tree Violation and Ambivalence in Virgil', *TAPA*, 118: 261–73.

Thommen, L. (2012), *An Environmental History of Ancient Greece and Rome*, trans. P. Hill, Cambridge: Cambridge University Press.

References

Tucker, G. M. (1997), 'Rain on a Land Where No One Lives: The Hebrew Bible on the Environment', *JBL*, 116 (1): 3–17.

Turcan, R. (2000), *The Gods of Ancient Rome*, trans. A. Nevill, Edinburgh: Edinburgh University Press.

Tylor, E. B. (1871), *Primitive Culture: Researches into the Development of Mythology, Philosophy, Religion, Art and Custom*, vol. 2, London: John Murray.

Vögler, G. (1997), *Öko-Griechen und grüne Römer?*, Dusseldorf: Artemis & Winkler.

Warrior, V. (2006), *Roman Religion*, Cambridge: Cambridge University Press.

Weeber, K. -W. (1990), *Smog über Attika: Umweltverhalten im Altertum*, Zürich: Artemis.

Whaling, F. (1984), 'Comparative Approaches', in F. Whaling (ed.), *Contemporary Approaches to the Study of Religion*, vol. 1, 165–295, Berlin: Mouton.

White, L., Jr. (1967), 'The Historical Roots of our Ecologic Crisis', *Science*, 155: 1203–07.

Wright, G. E. (1952), *God Who Acts: Biblical Theology as Recital*, London: SCM Press.

York, M. (2003), *Pagan Theology: Paganism as a World Religion*, New York: New York University Press.

York, M. (2016), *Pagan Ethics: Paganism as a World Religion*, Cham: Springer.

INDEX

Achilles 34
Aeneas 26
Aeschylus 8, 11, 22, 32, 75, 103–18
    Oresteia 11–12, 103–18
    Persians 32, 112–13, 114–15
    Seven Against Thebes 113
aesthetics. *See* beauty
agency, human 13–16, 23
agriculture 15, 16, 18, 21, 22, 23, 28, 29, 103, 104, 113, 116, 119, 133
Aldington, Richard 133
altruism 68, 69, 73
Ambrose of Milan, St 79, 87
anachronism 6, 7, 124
Anaxagoras 25, 76
Anaximander 115
Anaximenes 76
Andromache 31
animals 5, 14, 17, 21, 27, 28, 30, 47, 49, 54, 55, 56–7, 67, 68, 70, 72, 74, 77, 78–9, 80, 81, 82, 83, 85, 86, 97–8, 99, 100, 109, 115, 116, 119, 125, 126
animism 12, 137–52
Anthropocene 1, 13, 14–15, 16, 23
anthropocentrism 10, 11, 18, 23, 30, 38, 49–50, 51–3, 56, 60, 65, 77, 79, 80–3, 91, 124, 137, 146, 147, 151
anthropomorphism 37, 144
Aphrodite 143
apocalypticism 9, 27, 36, 37, 38, 126–7, 132, 135. *See also* eschatology
Apollo 106
Aquinas, Thomas, St 83
archaeology 146
Argos 105, 117
aristocracy 23
Aristotle 11, 15, 45, 64, 83, 84, 88
    Eudemian Ethics 42
    Metaphysics 77, 80–1
    Meteorology 81, 85
    Nicomachean Ethics 42
    Parts of Animals 80
    Physics 42, 77
    Politics 77, 80, 81
Artemis 106–7
Arval grove 144
Asia Minor 27, 32

Assyria 35, 123
atheism 73–4, 87
Athena 22, 106–7, 108, 109, 117
Athenaeus 119, 144
Athens 18, 22, 106–9, 117
Augustine of Hippo, St 83, 86, 87
Augustus, emperor 37, 39
autochthony 9, 20, 22, 23
Axial Age 13
axiology 50, 54, 56, 57, 80, 82

Babel, Tower of 34, 35
Babylon 36, 100, 123, 126
Basil of Caesarea, St 79, 83, 86, 87
Bauckham, Richard 147–8, 150, 151
Beatniks 51
beauty 5, 10, 26, 28, 39, 45, 57, 58, 59, 63–74, 77, 80, 84, 87, 96
Bellhouse, Stuart 129
Benedict, St 86, 88
Bible, Hebrew 8, 32, 37, 51–2, 78–9, 83, 86, 88, 91–101, 127, 128, 146
    Genesis 16, 17, 18, 27, 35, 82, 94, 95, 96, 126, 127, 137, 146–7
    Exodus 9, 19, 20, 28
    Leviticus 21, 31, 33
    Numbers 98
    Deuteronomy 18, 20–1, 31, 119, 124, 126, 134
    Joshua 93
    Job 11, 83, 93, 97, 98, 100–1
    Psalms 8, 11, 37, 78–9, 83, 92, 93, 94, 97, 101, 126, 147
    Proverbs 79, 83
    Ecclesiastes 99
    Isaiah 8, 28, 30, 35, 36, 37, 79, 123, 126
    Jeremiah 12, 123, 125, 126, 127, 129, 131, 132, 134
    Lamentations 31, 96, 124
    Ezekiel 11, 94, 95–6, 121
    Daniel 37
    Hosea 30, 35, 125, 131
    Joel 95, 124, 125, 126
    Amos 30, 33, 35, 125, 126
    Jonah (Jon.) 11, 92, 93, 94, 96
    Micah 126
    Nahum 123
    Habakkuk 99, 100

Index

Zephaniah 124, 126
Zechariah 95
2 Esdras 11, 95, 96
bio-climatology 75
biocentrism 49, 81, 85, 86
biodiversity 1, 7, 47, 61, 84, 88, 89
biology 77, 84
biosphere 13, 47, 61, 73, 84, 88
blood 31, 35, 92, 111, 112, 115, 116, 129, 130
Blunden, Edmund 131, 135
Bookchin, Murray 87
Borden, Mary 130, 132
Brand, Stewart 13
Brundtland Report 89
Buddhism 51, 148
Bunyan, John 135
Byng Scott, Aimee 129

Cain and Abel 16
Calvin, John 83
Canaan 122–3
carbon emissions 1, 13, 76
Carlyle, Thomas 5
Carson, Rachel 5
causation 26, 37, 64
childbirth 108, 124
Christianity 2, 6, 8, 9, 12, 14, 25, 26, 38, 51–2, 61, 79, 82–3, 85–6, 88, 126, 137–8, 139, 142, 145–52
Chrysippus 41, 49, 50, 55, 56–7, 58, 66, 77, 82
Chrysostom, John 79, 83, 86
chthonic deities 11, 103–18
Cicero 7, 49, 50, 52, 54–6, 66, 68–9, 71, 77, 81–2
Cistercians 88
Civil War, American 121
Cleanthes 58, 70
Clement of Alexandria, St 87
climate change 7, 14–15. *See also* global warming
comparativism 138, 139
contamination, environmental 1, 120, 129
Cosmas Indicopleustes 87
covenant 19, 31, 126
craftsmanship 66, 67, 71
creation 11, 12, 16–17, 18, 23, 29, 35, 37, 44, 45, 46, 52, 61, 64, 66, 67, 68, 78, 79, 82, 83, 87, 91, 92, 93, 94, 97, 100, 101, 124, 126–7, 135, 146, 147, 148–9, 151
Crete 27
Crutzen, Paul 1
cultural appropriation 4, 30, 122
Cumae 26, 28, 33
Cuthbert, St 83
cycles, agrarian 2
Cypria 35
Cyprus 36

daimones 17, 108, 110–11, 116
Darwinism 11, 76, 78, 88
de-creation 12, 126–7
Deep Ecology 49, 51, 61
deforestation 18, 84, 85, 119, 142. *See also* trees
Delphi 106, 117, 144
Demeter 103, 104, 133
Demiurge 44, 45, 76
democracy 14, 51
Descartes, René 52
design, intelligent 46, 64
Diodorus Siculus 119
Diogenes Laertius 58, 64, 82
Dionysus 103, 144
disease 21, 58, 108, 109
disorder, cosmic 12, 71, 126, 131
Dodona, oak of 144
dominion, human 49, 50, 52, 53, 93, 100, 137
Dorylaeum 30, 36
drought 2, 31, 125, 131
dualism 51, 52, 55

Earth Bible Project 124
earthquakes 36–7, 126
eco-capitalism 4
eco-feminism 4, 149
eco-modernism 4
ecocentrism. *See* biocentrism; nonanthropocentrism
ecocriticism 4, 10, 26, 38–9, 137, 145, 151
ecology 2–6, 9–12, 13–20, 22, 23, 25–7, 29, 31–2, 38–9, 49, 51, 53, 54, 61, 73, 75, 76, 77, 78, 79, 83, 84, 85, 87, 88, 91, 101, 103, 104, 117, 120, 122, 124, 125, 135, 136, 137, 138, 141, 142, 149, 150–2
economics 5
ecosophy. *See* Deep Ecology
ecotheology 3–4, 6, 12, 103, 105, 117, 137, 138, 146, 149, 151–2
Eden, Garden of 5, 18, 21, 96
Egypt 20, 27, 33, 36, 37, 93, 122
  plagues of 9, 19
Empedocles 10, 42–5, 76
environmentalism 1, 5, 50, 53, 60, 75–6, 78, 79, 83–4, 85, 87, 88, 89
epic 9, 25, 29, 31, 33, 35, 37, 75, 88, 105, 115, 121
Epictetus 68, 72
Epicureanism 78
Erinyes 11, 105–10, 112, 115–18
eschatology 25, 27, 28, 31–2, 37, 57. *See also* apocalypticism
ether 37, 66, 71
ethics 1, 3, 5, 8, 10–11, 21, 22, 25, 32, 39, 63, 69, 73, 74, 76, 77, 84, 99, 124, 137, 138, 142, 145, 147, 149, 150, 151. *See also* virtue ethics

Ethiopia 36
Etruscans 26
eudaemonism 58, 63
eudaimonia. See eudaemonism
exegesis 8, 35
extinction 13, 15, 127

Faunus 143–4
fertility 20, 22, 52, 86, 88, 103, 104, 106, 108, 109, 117, 122, 124
Fichte, Johann Gottlieb 78, 87
First World War 12, 119–36
fishing 1, 7, 91, 96, 98–100, 101
flood 2, 27, 30, 34–5, 36, 95, 126
flourishing, human 11, 31, 60, 63, 68, 69, 72–3
Francis of Assisi, St 14, 51, 83, 88
Francis, Pope 1–2
fruit 16, 17, 18, 33, 55, 81, 82, 86
Furies 22

Gaia hypothesis 10, 43, 61, 76, 84
Gaul 27
generative thesis 137
genii 138, 140
geology 1, 13, 14–15, 16, 23, 67
Gibraltar, Straits of 96
gift, earth as 28, 124, 136
global warming 84, 88, 89, 93. *See also* climate change
Gnosticism 78, 150
golden age 16, 17, 26, 28, 29, 32–3, 34, 134
Golden Rule 5
Golding, William 84
Gottlieb, Roger 2, 5
groves, sacred 141–4, 146

Hades 95, 104
Haeckel, Ernst 5
Hale, Matthew 86
hamadryads 143
harvest 2, 32, 72
    as metaphor 113–15
Hasler, Digby B. 131
Hecate 104
Hecato 58
Hector 31
Hegel, Wilhelm Friedrich 87, 146
Helen of Troy 30, 116
heliocentrism 42
henotheism 10, 50, 54
Hephaestus 22
Heraclitus 115
hermeneutics 2, 6, 122
Hermetic Corpus 78, 87–8
Herodotus 75, 119

Hesiod 15, 25, 26, 27, 38, 115
    Theogony 105
    Works and Days 16, 17, 22, 33, 105
Hinduism 148
Hippocrates 75
history, views of 14–15, 23, 26–7, 29–30, 33, 38, 57, 61, 137, 142, 146, 147, 148
Holocene 1, 13
Homer 8, 25, 26, 32, 36, 38, 80, 135, 143
    Iliad 31, 34, 37
    Odyssey 27, 33, 144
Housman, A. E. 132
hubris 11, 22, 26, 28, 50, 56, 75, 91, 94, 96, 101, 108, 110, 111, 113
hunting 15, 55, 99

immortality 46
imperialism 110, 113, 115, 139
industrialization 2, 5
Industrial Revolution 13, 88
intertextuality 8
Irenaeus 79
Islam 79
Israel, kingdom of 122, 125

Jerusalem 31, 32, 36, 95, 124
    temple of 36, 95–6
jubilee year 21–2
Judah, kingdom of 100, 123, 125
Judaism 8, 28, 30, 31, 35, 36, 37, 83, 86, 126, 146, 150
    Hellenistic 9, 25–6, 38
judgement, divine 27, 35, 123, 125, 126, 135
Jupiter 65, 144
justice 17, 22, 33, 39, 54, 56, 57, 60, 105–7, 110, 111, 115, 116, 117

kingship, divine 17, 33, 165 n.54
Krakatoa 84
Kronos 16–17, 18, 30, 33
Kyoto Protocol 7

labour 9, 13, 16, 18, 21, 105, 111, 124
landscape 6, 26, 27, 87, 113, 120, 121, 123, 125–31, 133, 134–6, 140
Leopold, Aldo 5, 76–7, 84
Leviathan 83, 97–8
Libya 27
logocentrism 53, 56, 81
Lovelock, James 43, 76, 84, 148
Lucretius 76, 78
Lycia 34
Lydia 144

Macedonian wars 31
Marcus Aurelius 41, 46–7

Index

McFague, Sallie 148–51, 152
Menalcas 39
merisms 27, 132
Mesopotamia 122
Messianic prophecies 26
meta-ethics 59
metaphors, environmental 10, 26, 29–32, 36, 37, 39, 87, 99, 100, 115, 116
metics 22
metre 25, 38, 123
Mew, Charlotte 134
Midgley, Mary 77
Millennium Development Goals 11, 87, 89
Milton, John 135
Minerva 144
mining 18
Moltmann, Jürgen 146, 148–52
monotheism 13–14, 148, 150
Moses 19, 98
mother, earth as 22–3, 75, 113, 125, 129, 133–4
mourning 12, 35, 124–6, 129, 131
Muir, John 5
mythology 7, 26, 29–30, 36, 97, 127, 135, 143, 145

nature 1–6, 9, 11, 12, 13–16, 18, 20, 23, 25, 28, 30, 33, 35–7, 49–53, 55, 57–61, 65–7, 69, 76, 77–8, 79, 80–3, 85–8, 93, 103, 105, 107, 109, 111, 115–17, 126, 128–9, 134, 137, 138, 140, 141, 144, 145–51
neopaganism 150–2
Neoplatonism 78, 88
Neo-Stoicism 49, 50, 85
Nero, emperor 31, 34
New Testament 8, 79, 83, 86, 88
    Matthew 32, 37, 79, 83, 86, 126
    Mark 83, 86
    Luke 86, 126
    Romans 79, 83
    Revelation 37
Newton, Isaac 148
Nichols, Robert 128
Noah 26, 27, 34
nonanthropocentrism 60
numen 139–40, 143–4
Nussbaum, Martha 73, 77, 80–1, 85
nymphs 143

Octavian. *See* Augustus, emperor
Old Testament. *See* Bible, Hebrew
Oracle of the Potter 33–4
oracles 9–10, 25–39, 94, 96, 106, 121, 123, 126, 145
order, cosmic 12, 63–4, 71, 74, 76, 78, 105–8, 112–13, 115, 116–17
Origen 11, 52, 79

Orpheus, Testament of 37
Ovid 6, 29, 143–4

Panaetius 47
panentheism 148–9
panpsychism 147
pantheism 10, 50, 54, 55, 146, 148–50
Passmore, John 11, 51–3, 76, 77–8, 79, 80, 81, 82, 83, 85–7, 88
pastoral tradition 28, 128, 134
Paul, St 6, 25, 52, 79, 83, 149
Pausanias 145
Pemberton, Vivian T. 129–30
Pepi I, pharaoh 122
perichoresis 149
Persephone 103, 104
personification 12, 124, 126, 129, 133, 135. *See also* mother, earth as
philosophy, environmental 1, 3, 5, 6, 8–9, 10–11, 14, 41–7, 49, 51, 52, 56, 59, 75–89, 104–5, 115, 137, 145
Phoenicians 94
Phrygia 30
Picus 143
plastics 1, 7, 59
Plato 8, 15, 19, 20, 64
    Critias 18, 22
    Euthyphro 41
    Gorgias 77
    Laws 16, 17, 18
    Menexenus 23
    Phaedo 76, 77
    Phaedrus 78, 84, 87, 88
    Republic 7, 14, 22, 76
    Statesman 16, 17
    Timaeus 10, 11, 42, 44–7, 66, 76, 78, 83–4, 88
Pliny the Elder 23, 143–4
Plotinus 70, 71
Plutarch 56–7, 145
Pluto 104
pneuma 64
poetry 12, 25, 26–7, 29, 30, 33, 38, 39, 42, 75, 80, 88, 105, 121–36
polis, the 22, 51, 67, 73, 117, 139–40
politics 1, 4, 9, 13–23, 39, 100, 109, 120, 122, 123, 125
polytheism 10, 14, 50, 54, 56, 57, 146, 150, 151
Porphyry 78, 88
portents 37, 38, 39, 144
Poseidon 36
Posidonius 51, 77–8, 87, 88
poverty 58, 119
prehistory 25
Presbyterianism 5
Presocratics 76, 80

198

Index

promised land 123
prophecy 9, 25, 26, 27, 28, 30, 31–2, 33, 35, 36, 37, 38. *See also* Messianic prophecies; oracles
providence 56–7, 63, 65–6, 68–72
Purdy, Jedediah 1

Quellenforschung 139

rationality 52–3, 63, 64, 65, 66, 67–8, 69, 70, 72, 74, 79, 147
recycling 10, 41, 46–7
Red Sea 20
reformation, Protestant 146, 148
religious adherence 1, 52
Rhea 30
rhetoric 27, 54, 98, 126, 127, 132
Rio Earth Summit 89
Rollston III, Holmes 6
Rome 25, 32, 144
Romulus and Remus 144
Rosenberg, Isaac 133–4, 135
Rousseau, Jean-Jacques 60
Ruminal fig 144
Ruskin, John 32

sabbatical year 21–2
sagas, Norse 121
Salamis, battle of 113
Sardinia 36
Sassoon, Siegfried 127–8
Saunders, Kenneth 130
scala naturae 64, 66, 72
Scamander, river 34
Scotland 5
sea 11, 18, 27, 29, 34–5, 36, 42, 43, 44, 47, 49, 52, 59, 60, 69, 71, 79, 85, 88, 91–101, 108, 111, 113, 115, 117, 120, 122, 134
sea level 15
self-sufficiency 10, 41–7, 64, 65
Sen, Amartya 73
Seneca 49, 59–60, 61, 65, 67, 71, 72, 85
Servius 34, 143
Sheldrake, Rupert 147–8
Sibyl 9–10, 25–39
Silenus 29
Silius Italicus 144
sociability 10, 63, 64–5, 68, 69, 70, 72, 73–4
Socrates 41, 76, 77
soil erosion 18, 120
Solon 115
Sophocles 75, 80
Sparta 31, 144
speciesism 56
Speth, Gus 7

Sphairos 10, 43–4, 45
Spinoza, Baruch 49, 60, 148
spirituality 7, 127, 146
Spoleto, Italy 144
stewardship 7, 11, 18, 23, 76, 79, 86–7, 88, 91, 100
Stoicism 10–11, 26, 41, 42, 47, 49–61, 63–74, 77–8, 80, 81–3, 85, 87, 88. *See also* Neo-Stoicism
storms 19, 27, 31, 34–6, 52, 92, 93, 108–9, 115–16
Strong, Maurice 7
superorganisms 10, 43, 45, 47, 84
Sustainable Development Goals 11, 87–8, 89
sustainability 1, 4, 7, 11, 14, 16–17, 18, 21, 46, 69, 87, 89, 98
sympathy, cosmic 32, 50–1, 70–1

technology 1, 2, 7, 14, 49, 51, 52, 53, 61, 82, 84, 93, 101, 137, 150
teleology 14, 52, 55, 56, 57, 68, 71, 73, 74, 76, 78, 81–2, 85
temperance 60
theism 11, 63, 65, 73, 74
theocentrism 147
Theodoret 79, 87
theology 2, 3, 6, 8, 9, 10, 12, 13–23, 37, 38, 47, 49–50, 54, 56, 57, 61, 66, 75, 79, 93, 94, 103, 105, 106, 108, 135, 136, 137, 138, 142, 145, 146, 147, 148, 149, 152. *See also* ecotheology
theophany 37
Theophrastus 7, 77, 88
Thessaly 7
Thoreau, Henry David 5
thought experiments 46
Thucydides 119
Tiglath Pileser III, king of Assyria 123
Titans 30, 34
Tomlinson, Albert E. 131
trauma 121, 123, 127, 132
trees 16, 17, 18, 20–1, 33, 54, 86, 119, 129, 134, 141, 142–5, 150. *See also* deforestation
Trinity, Holy 148–9
Trojan War 30, 35
Troy 27, 30, 36, 38, 105–6, 110, 115, 116, 117
Tylor, Edward Burnett 139
Tyre 94–6

Ugaritic 95
underworld 11, 26, 94, 95, 96, 103, 104
Uni, Egyptian general 122
United Nations 87, 89
utilitarianism 63, 71, 72, 73, 98, 101

vegetarianism 78, 88
Verdun, France 120
Vesuvius, Mount 32, 36

Index

Virgil
  Aeneid 143
  Eclogues 10, 26, 28–9, 32–4, 37, 39
virtue ethics 42, 49–61, 73, 85

war 12, 17, 18, 21, 22, 29, 31, 34, 36, 39, 111, 119–36
wealth 12, 58, 105, 107, 108, 109, 110–18
weather 19, 27, 30, 31, 36, 37, 115

Wheeler, William Morton 43
White, Lynn, Jr. 12, 14, 51, 82, 83, 85, 88, 137–9, 140, 141–2, 145–7, 150–1, 152
wildlife 1, 119, 120, 131
Wordsworth, William 121
World War One. *See* First World War

Zeno 58, 66
Zeus 30, 33, 35, 58, 70, 103, 106, 107, 108, 112, 114, 144